HOT ROD
volume 6
TECHNICAL LIBRARY

The Best of
HOT ROD
MAGAZINE

High Performance Small Block FORD Engines

CarTech
Auto Books & Manuals

Published by CarTech, Inc.
11481 Kost Dam Road
North Branch, MN 55056 U.S.A.
United States of America
Tel: 800-551-4754, fax 651-583-2023

ISBN 1-884089-37-2

Book Trade Distribution by
Voyageur Press, Inc.
123 North Second Street
P.O. Box 338
Stillwater, MN 55082 U.S.A.
Tel: 612-430-2210, fax 612-430-2211

Distributed in England by
Brooklands Books Ltd
P.O. Box 146, Cobham
Surrey KT11 1LG
England
Tel: 01932 865051, fax 01932 868803

Distributed in Australia by
Brooklands Books Ltd
1/81 Darley St, P.O. Box 199
Mona Vale, NSW 2103
Australia
Tel: 02 9997 8428, fax 02 9979 5799

Printed in Hong Kong

CONTENTS

THE 351 PART 1 FORD SHOULD HAVE BUILT

The core of the conversion is the modified 400ci connecting rod, shown here next to a stock 351W. The Wayne's Engines rods are void of balance pads, have been bushed for full-floating wristpins, and are equipped with high-strength ARP bolts.

The path to 11:1 compression, 87-octane gas, and 400 hp with factory EFI

text and photography by Pete Saueracker

Editor's note: A little more than a year ago, Pete Saueracker stirred up a hornet's nest with a long-rod 350 Chevy engine buildup in Hot Rod *magazine. The engine made huge power on low-grade 87-octane gas, and Saueracker attributed it all to the theory of long-rod versus short-rod technology.* Hot Rod *was deluged with letters calling the story everything from a godsend to utter fabrication, so we enlisted Pete to prove it again, this time with a fuel-injected 351. This story is part one, and it shows the combination and discusses the theory. Next month, we'll bring you part two, with more extensive dyno tests.*

Those of us old enough to remember the horsepower of the '60s often dream of feeling that thrill of brute power again in the current environment of stringent emissions and high fuel-mileage requirements. But no matter what age you are, few seriously think that a factory fuel-injected V-8 will ever produce the massive horsepower and torque that marked the golden age of musclecars.

Although the Big Three focused on fuel mileage and low emissions nearly 25 years ago, small groups of performance junkies have been applying the latest advances in engine science in ways their originators never dreamed. We are among that group, and by employing these ideas and concepts, there is a way to return '60s acceleration and power, yet maintain the superior driveability, emissions, and mileage offered by factory fuel-injected cars.

If You Could Have It All

Imagine a fuel-injected engine that uses the factory computer and maintains all its mandated strategies. Then combine it with 11:1 compression, inexpensive 87-octane gasoline, and performance that catapults beyond 400 hp! Sweeten the mix with the knowledge that all of this can be achieved by the backyard hot-rodder through bolt-on parts. A laptop computer isn't needed to reprogram the fuel-injection system, and the basic long-block can be assembled for $3,000-$4,000.

The key to the jump into hyperspace is the realization that traditional hot-rodding techniques have focused on getting more air and fuel in and out of the engine, but have largely ignored what happens inside the combustion chambers once the valves have closed. By understanding and improving the process of combustion itself, significant performance increases result, no matter how much you have already hot-rodded your engine with traditional methods. Best of all, when you improve the combustion process, you gain the significant benefit of heightened efficiency, which translates into lower emissions, higher fuel mileage, and a reduced sensitivity to fuel detonation.

There are three secrets to achieving 400-plus horsepower in a modern fuel-injected V-8, and the key to it is longer-than-normal connecting rods. To complement the special rods, spice the mixture with very high compression, and flavor it with dishtop pistons to promote combustion efficiency. Let's look at what each of these three secrets has to offer.

The Long and The Short of It

Increasing the length of the connecting rod beyond the stock dimension improves combustion efficiency by extracting more power out of the same amount of fuel and air. The physics of this gets a little complicated, but much of it has to do with the position of the piston relative to the combustion chamber as the crankshaft rotates. The longer the connecting rod, the longer it takes the

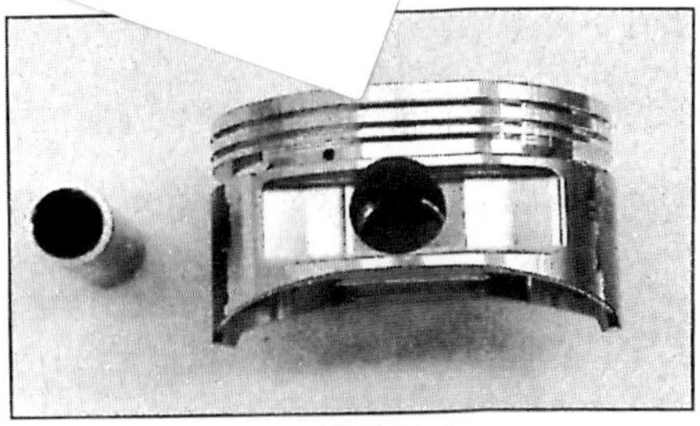

The wristpin for the matching JE pistons is situated just below the oil ring. The forging is a paltry 415 grams, and uses an .866-inch–diameter, 77-gram wristpin designed for a Porsche. The dished recess in dome (darkened area) matches the combustion chamber of the Trick Flow cylinder heads.

A two-bolt main 351W block does just fine, thank you. The holes in the crankshaft's rear counterweight compensate for the lightness of the special pistons and rods, all of which contribute to a free-revving engine.

piston to drop away from the top of the cylinder bore (top dead center, or TDC) on the combustion stroke. Because the piston stays near the top of the cylinder longer as the crankshaft spins, the volume of the chamber remains smaller during the initial combustion process than it would if the rods were shorter. By keeping the combustion chamber smaller in volume for a longer period of time, average combustion pressure remains higher and thus more work is extracted from the burning air and fuel to push the piston down. This enhances the power curve.

The 351W connecting rod is 5.956 inches long, while the 400ci rods used

A hydraulic roller cam (left) has extreme valve lift and short duration numbers. See how aggressive the roller lobes look compared to a standard hydraulic flat-tappet cam (right)? It was installed with a Crane double-roller timing set.

We began dyno-testing with the Trick Flow Twisted Wedge head, then switched to the TW R (racing) version. The TW heads flowed 250 cfm out of the box, while the R head flowed 290 cfm without any porting. Combustion chambers in both castings are identical.

for this project measure 6.580 inches, an increase of 0.62 inch overall. The L/R ratio of the 351W rod is 1.70:1, or about the same as that of a small-block Chevrolet. The 400 rod, however, offers an L/R of 1.88:1, or an increase of 10.6 percent.

Engine efficiency is defined as the ratio of potential power sent into the engine compared to the power sent out. The typical hot-rodded V-8 claims an efficiency of approximately 30 percent. That is, 70 percent of the energy is never used to turn the rear tires. Instead, it goes out the tailpipe as heat or unburned fuel (emissions), or is delivered to the engine

block where the coolant takes it away and uses it to heat the air. Because long connecting rods get more power out of the same amount of air and fuel, they increase overall efficiency on the order of 5 to 10 percent as delivered to the rear wheels.

When combustion efficiency is improved, sensitivity to detonation is simultaneously muted. Better detonation performance means higher compression without trouble. If you lengthen the rods enough, you can increase the static-compression ratio of your engine, and reduce the octane rating of the gasoline at the same time. Increasing the static-compression ratio is another step that promotes improved efficiency and an increase in power. The use of low-octane gasoline saves in the operating costs of your engine, leaving more money to spend on other improvements, or the difference between having a daily driver rather than a trailer queen.

The benefits of long rods are enhanced by the use of dished piston tops, which get more of the combustion process to take place "in" the piston, rather than along the roof of the combustion chamber. The shape of the combustion chamber in any engine is defined by the walls of the cylinder heads, the cylinder walls, and the configuration of the piston top. These metal surfaces operate at a colder temperature than that of the burning fuel and air (they have to or they would melt), and therefore suck heat out of the combustion process. This reduces the power of the expanding gases as they work on the piston to push it down in the cylinder. The ideal combustion chamber would be a sphere, which has the least amount of wall area for any given volume. In engineering terminology, the sphere has the smallest surface-to-volume relationship of any shape. Using dished pistons makes a typical wedge combustion chamber more spherical, and gets more of the combustion process to take place in the center of the combustion chamber. Less heat is lost to the walls, and more is available to make power at the crankshaft. This is an extremely simplified version of what actually goes on in a long-rod engine, but at least it gives you the basics.

Applying Theory To Reality

The benefits of long rods and high compression apply to any engine, no

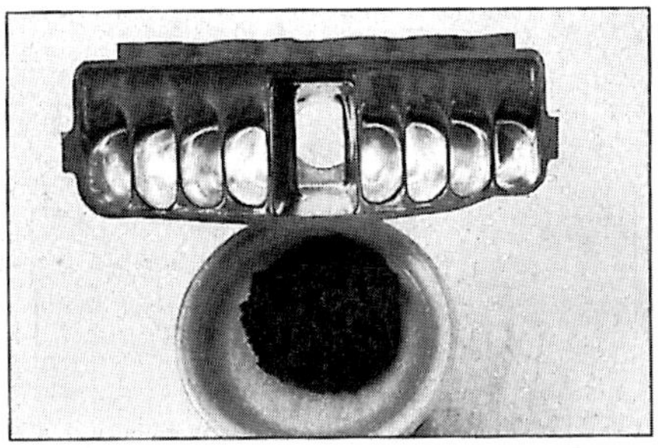

Factory fuel-injection systems suffer from insufficient airflow, so Extrude-Hone opened up our intake manifolds with the company's thick, sandy "putty" (in bowl). We gained about 70 cfm by this method, raising airflow from 220 to approximately 290 cfm per runner.

Aluminum heads need guideplates to locate hardened steel pushrods. ARP fasteners throughout squeeze Fel-Pro performance gaskets. The rings are pre-gapped Speed Pro (Dana) moly units.

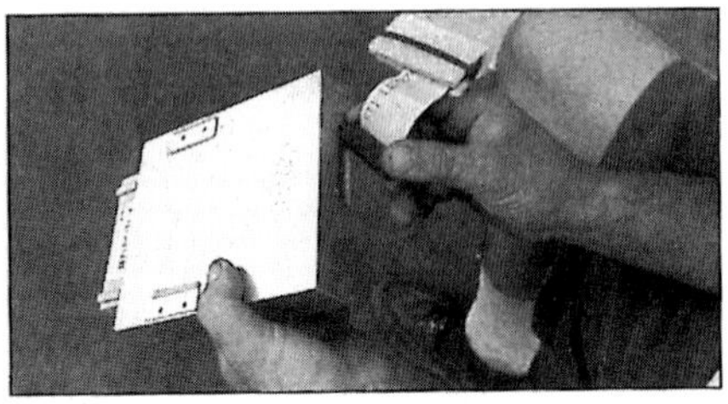

To modify the EEC-IV computer, unscrew the cover and clean the terminals for the connector at the back of the unit. Reinstall the cover and plug in the special custom Kenne-Bell chip.

matter what the brand or type. You can make these theories work on a '21 Stutz, if that's your preference. We, however, decided to build a long-rod Ford 351W, an engine that presents extra engineering challenges compared to building a 5.0.

According to Ken Duttweiler—builder of mondo Buick turbos and Hurley Blakeney's sub-7-second turbo-charged Ford small-blocks—the 351W suffers intake breathing problems that

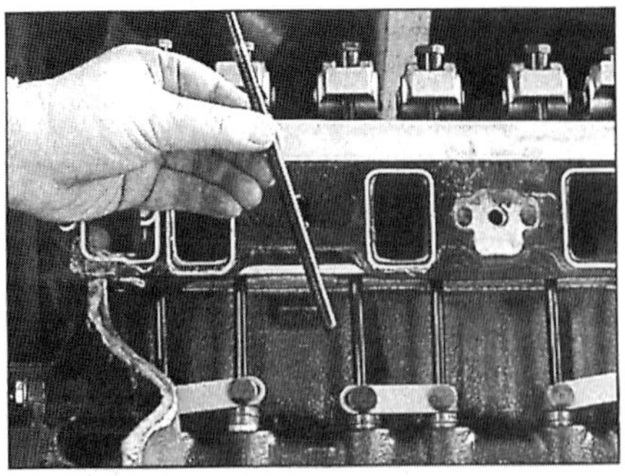

To obtain the desired 11:1 compression ratio, the heads are milled .070 inch to reduce combustion-chamber volume. This requires custom-length pushrods to reinstate proper valvetrain geometry. The TW head has 2.02-inch intake and 1.60-inch exhaust valves. The R head uses .300-inch longer valves combined with 2.080-inch intakes and 1.60-inch exhausts.

We tested an Edelbrock Performer 5.0 intake manifold and the Ford SVO Cobra manifold. The Cobra piece gave us more low-end torque, while the Performer produced the most peak torque and horsepower. So that the upper 5.0 Mustang half of the two-piece manifold would clear the air-charge temperature sensor in the lower half of the 351W truck manifold, we had to use a pair of phenolic-resin spacers from BBK Performance. These spacers stifle heat transfer from the engine block to the upper manifold.

The Society of Automotive Engineers (SAE) has established the test procedure that the Big Three follow when dyno-testing and rating their engines. To comply with the SAE procedure (called SAE J-1349, net), you must employ the intake system, accessories, and exhaust as used on the car, and with full emissions equipment in place. We fabricated an aluminum intake plenum box to which air from Duttweiler's dyno was ducted, and from which our 351W engine drew its breath through the stock air-filter housing, filter element, and intake ducting. It made our testing more complicated, and probably reduced peak power and torque numbers, but made the data fully compatible with the numbers the OEMs generate in their own testing.

the 5.0 Ford engine does not. The 35 engine is taller than the 5.0/302 engine that it is based on. The problem arises when fitting the taller engine under the hood of a Mustang that was designed for the shorter 302. The factory flattens down the intake manifold, and introduces some sharp curves to the intake runners, which hinder airflow at high rpm. When aftermarket companies started equipping the 351W, they were forced to maintain Ford's flattened-out manifold dimensions, or their parts wouldn't fit under the stock hood. As a result, it is harder to get the 351W to breathe well, and thus harder for it to make power. It is a perfect candidate for the application of our long-rod theories.

EFI and Power

As if the 351W's restricted breathing was not challenging enough, we wanted to use fuel injection instead of carburetion. Under carbureted conditions, delivering sufficient airflow to the 351W to support in excess of 450 hp would be easy. However, electronic fuel injection (EFI) is notorious for falling way short of carburetors in the power department. Many feel that EFI simply can't match a carb for peak power output, but don't fully understand why this is so (we'll explain it later). How would we be able to reach the desired power level on unleaded regular gas and the Ford EEC-IV EFI system when the very best any OEM has been able to achieve is 345 hp on premium fuel (the C-5 Corvette)?

The answer lies in knowing that EFI isn't the problem in making power and torque, but that airflow is. EFI has all the abilities of a carburetor when it comes to making peak power and torque, and offers tons more regarding driveability, cold startup, and the like. The problem is that all factory EFI manifolds simply cannot pass enough air at the upper-rpm ranges to supply the engine adequately. This knowledge helped us when we formed the engine parameters and selected the parts.

Preliminary Results

We built our 351W using two sets of aftermarket cylinder heads from Trick Flow Specialties. Check out the accompanying photos for an overview of the parts and assembly procedures used for this engine. Our first session lasted four days at Duttweiler's, whereupon we swapped heads and rode the dyno twice more. The results of the first testing session are presented here. The results of the last two tests, along with more technical background information, will be presented next month in the second part of this series.

THE 351 FORD SHOULD HAVE BUILT

How'd we do? Our first dyno session yielded a completely emissions-legal, 87-octane V-8 that put out 360 to 380–plus hp, and torque in the 420 to 430 lb-ft range! This matches all the '60s small-blocks, surpasses the LS1-powered Corvette, and costs a whole lot less. Stay tuned!

Long-Block Parts

Part	Description	Source
Block:	Stock two-bolt 351W, bored .030-inch oversize	Local
Crank:	Stock 351W crank modified for long rods	Wayne's Engines
Rods:	Ford 400-cid modified for .866-inch floating pins and 351W crank, with ARP rod bolts	Wayne's Engines
Bearings:	.010-inch oversize main (PN MS-1432), .030-inch oversize rod bearings (PN CB-927)	AE Clevite
Pistons:	415-gram dished forging with .866-inch pin (PN 139105)	JE Pistons
Rings:	Pregapped .030-inch oversize moly (PN R-9902)	Speed Pro (Dana)
Gaskets:	Overhaul kit with no retorque head gasket (KS2329)	Fel-Pro
Cam and lifters:	Hydraulic roller cam, 220-degree duration (PN 444131), hydraulic retrofit roller lifters (PN 36532)	Crane Cams
Pushrods:	7.635-inch hardened steel for standard head (PN 11636), 7.335-inch hardened steel for R head (PN 27638)	Crane Cams
Rockers:	1.7:1 aluminum needle bearing rollers (PN 36757)	Crane Cams
Rocker studs:	7/16-inch 20 top, 7/16-inch 14 bottom screw-in (PN 99151)	Crane Cams
Oil pump:	High-volume (PN 224-102R)	Speed Pro
Heads:	Twisted Wedge, 2.02-inch/1.60-inch valves (PN TFS-51400002), Twisted Wedge R head, 2.080-inch/1.60-inch valves (PN TFS-52400002)	Summit Racing

The entire long-block can be assembled for about $3,000 if you use the standard twisted-wedge small block Ford head from Summit. The basic long-block price increases to about $3,800 if you opt for the mammoth TW R heads. Compared to what you have to pay for a carbureted crate engine, this is still pretty inexpensive, considering the power and torque it can deliver with fuel injection in a fully emissions-compliant form.

Sources

AE Clevite Engine Parts
Dept. HR02,
325 E. Eisenhower Pkwy., Ste. 202,
Ann Arbor, MI 48108; (313) 663-6400

ARP Fasteners
Dept. HR02,
531 Spectrum Cir.,
Oxnard, CA 93030; (800) 826-3045

BBK Performance
Dept. HR02,
1611 Railroad Ave.,
Corona, CA 91720; (909) 735-2400

Chet Herbert Cams (JE Pistons)
Dept. HR02,
1933 S. Manchester,
Anaheim, CA 92802; (714) 750-1211

Crane Cams
Dept. HR02,
530 Fentress Blvd.,
Daytona Beach, FL 32114; (904) 252-1151

Perfect Circle, Sealed Power Division (Dana Corp.)
Dept. HR02,
P.O. Box 1208,
2001 Sandford St.,
Muskegon, MI 49443; (616) 722-1300

Duttweiler Performance
Dept. HR02,
1563 Los Angeles Ave.,
Ventura, CA 93004; (805) 659-3648

Edelbrock Corporation
Dept. HR02,
2700 California St.,
Torrance, CA 90503; (310) 781-2222

Extrude-Hone
Dept. HR02,
8800 Somerset Blvd.,
Paramount, CA 90723; (562) 531-2976

Fel-Pro Inc.
Dept. HR02,
7450 N. McCormick Blvd.,
Skokie, IL 60076; (847) 674-7700

Ford SVO
Dept. HR02,
44050 N. Groesbeck Hwy.,
Clinton Township, MI 48036-1108; (810) 468-1356

Kenne-Bell
Dept. HR02,
10743 Bell Ct.,
Rancho Cucamonga, CA 91730; (909) 941-6646

Trick Flow Specialties
Dept. 712011,
1248 Southeast Ave.,
Tallmadge, OH 44278; (330) 630-1555

Wayne's Engines
Dept. HR02,
2205 Business Way,
Riverside, CA 92501; (909) 684-2222

THE 351 FORD SHOULD HAVE BUILT PART 2

400 hp and 450 lb-ft with factory EFI and 87-octane gas

by Pete Saueracker

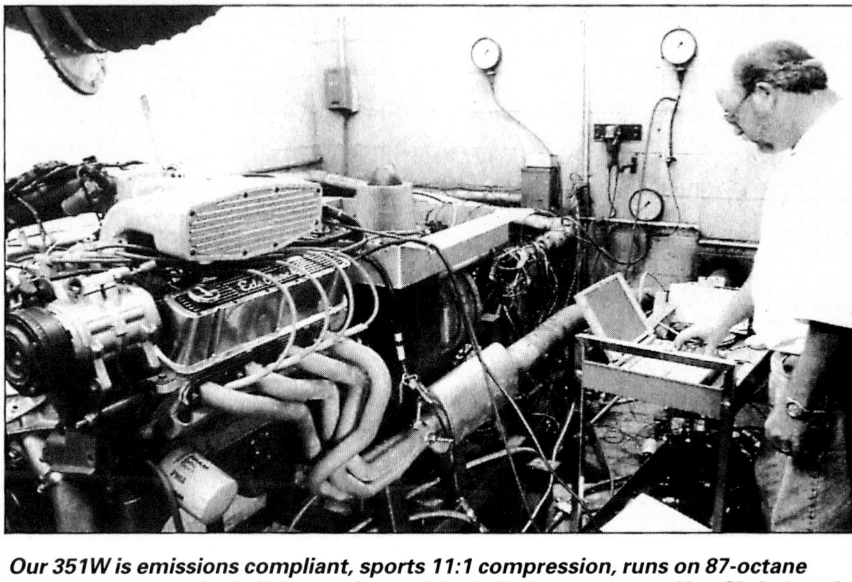

Our 351W is emissions compliant, sports 11:1 compression, runs on 87-octane unleaded, and can be built by any hot rodder in his home garage. Ken Christley of Kenne-Bell checks the A/F ratio as delivered by the company's computer chip.

Last month, we detailed the construction of a long-rod, high-compression Ford 351W V-8. Since this was a street-only engine, we purposely set our engineering goals very high. We required full emissions compliance, a factory-quality idle, and the inclusion of the Ford EEC-IV factory fuel-injection system to produce at least 400 hp and 400 lb-ft of torque. Further, we wanted to achieve peak power at 5,000 rpm and peak torque at 4,000 rpm.

Parts and Design

The basic engine theory and components were detailed in the first installment of this two-part article, but we'll briefly restate the concept here. The key to our engine is unusually long (6.58-inch) connecting rods, 11.0:1 static-compression ratio, optimized fuel delivery and spark timing, and wise selection and matching of the primary parts (see list). In a nutshell, longer rods allow the engine to make more power and to reduce its breathing needs, as well as its sensitivity to detonation, thereby allowing a relatively high static-compression ratio. A higher compression ratio elevates power and torque, and the reduced breathing relies on the long connecting rods to work better with OEM-type, air-limited intake systems and yield more go at low rpm levels.

The Problem with Factory Fuel Injection

The OE fuel injection posed three major engineering challenges: air, spark, and fuel delivery. First, let's look at providing optimum fuel delivery throughout the rpm band during wide-open throttle (WOT) and balancing it with the right amount of spark timing. Though the Ford EEC-IV mass-air SEFI system is adaptable, in stock form it cannot make the substantial changes necessary to provide the optimum air/fuel ratio for a 400hp engine such as ours. Further, the EEC-IV system's WOT preprogrammed spark curve is non-adaptable. Without addressing these challenges, our highly modified 351W wouldn't perform at full potential with stock EFI.

We found part of the solution at Kenne-Bell, in Rancho Cucamonga, California. The company supplied one of the few aftermarket chips available that can make the kinds of WOT fuel- and spark-timing changes needed by this offbeat engine combination. Traditionally, you couldn't adjust these factors with

Offsetting the wristpins 0.090 inch yields a slight angle to the rod when the crank is at top dead center (TDC); this effectively eliminates the deadlock between the two as the piston starts to push downward. This small amount of rod angle at TDC helps push the crank over more easily and should add low rpm power and torque. The penalty is a little more cold-start piston noise, but there is no increase in cylinder-wall wear.

"black box" factory EFI systems, but Kenne-Bell got into the factory computer and modified the WOT spark and the injector pulse control curves to deliver exactly what the motor needed. Since the company was involved in the development of this engine, it can supply the same chip we used.

In the matter of air delivery, even some aftermarket fuel-injection manifolds cannot pass the same amount of air that a good dual-plane carburetor manifold can deliver. To make matters worse, (according to engine-wizard Ken Duttweiler) the Ford 351W doesn't breathe as well as the Chevy 350, even in carbureted trim. Knowing this beforehand prompted us to enhance intake breathing from the start. An Extrude-Honed Edelbrock intake manifold, a 75mm throttle body, and an 85mm mass-air meter were all part of the path to power.

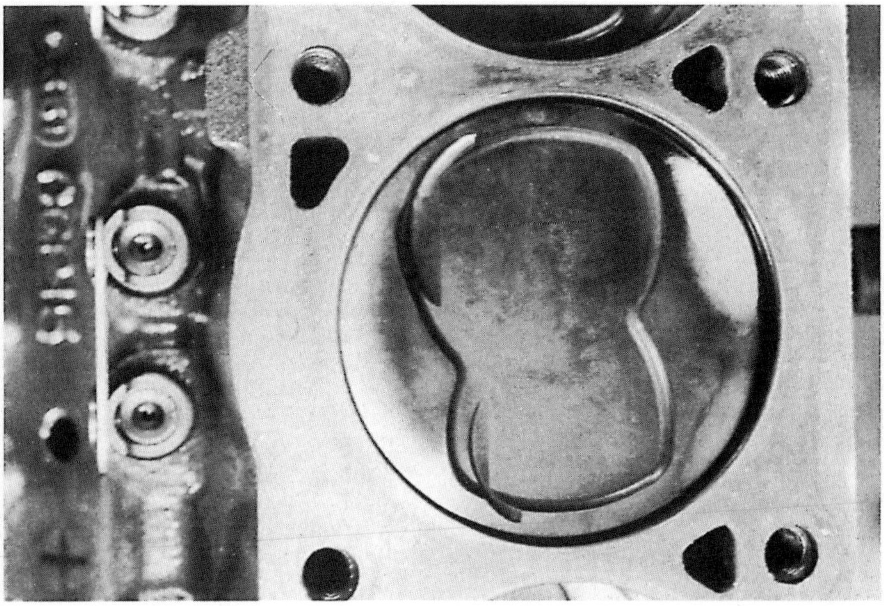

Part of the way through the testing, we pulled the heads to ascertain the flame-travel pattern. Reading the tops of the pistons helped to determine how much combustion heat was being made and how efficient the chambers were. (The bigger the flame pattern across the top of the piston, the more efficient your engine.) Our chambers looked very good, indeed.

Cylinder Heads

We tested with two different sets of heads on our long-rod 351W. The regular TFS Twisted-Wedge heads flow 250 cfm of air, as determined by our own flow-bench testing at Duttweiler's. Although this is a good number, we also dyno-tested our prodigious 351W with the TFS "R" head, which passed a whopping 290 cfm. Normally, such a large head would give up tons of low rpm torque, but by using long rods, the normal way of thinking doesn't apply anymore. In fact, the bigger heads produced 20 hp more at the 5,000-rpm peak and

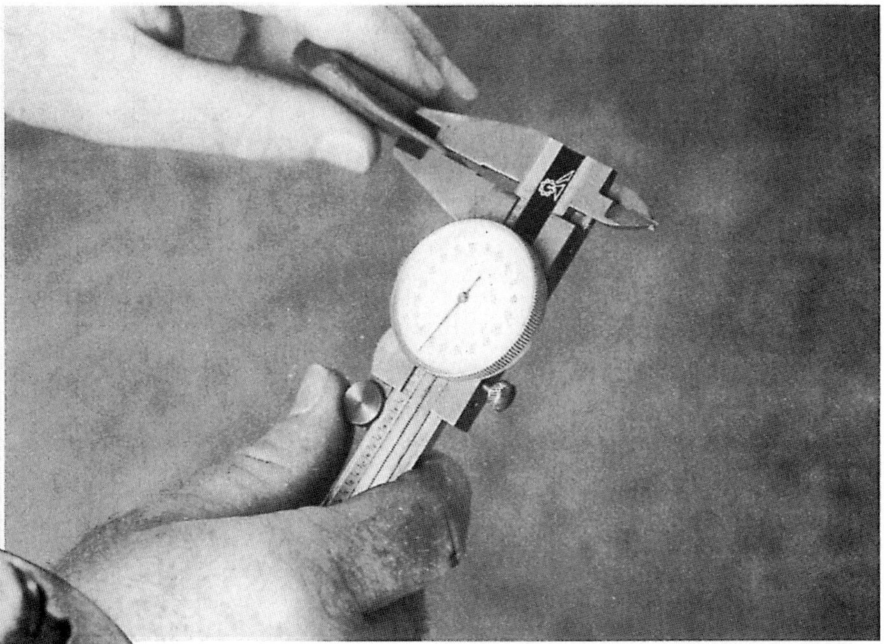

We double-checked piston-to-valve clearance by putting clay on top of the pistons during the mock-up process and rotating the engine through a couple of cycles. Our design yields 0.125-inch clearance on the intake side, 0.235-inch clearance on the exhaust side, and about 0.055-inch squish clearance between the flat portion of the piston and the bottom of the heads. The small squish clearance (being measured here) improves combustion-chamber efficiency.

A lot of low-rpm torque creates a lot of heat and pressure. While testing the limits of our design, we blew two pistons because an intake-valve cutout was made too deep, leaving only a few thousandths of an inch between the bottom of the intake-valve pocket and the back of the top ring groove. We then reduced the depth of the intake-valve cutout to yield 0.125-inch piston–to–intake-valve clearance. If you need a deeper valve cutout for a more aggressive camshaft, have the ring stack dropped down on the piston. Discuss these specifications with JE Pistons if you do not use the same cam we did.

20-plus lb-ft of additional torque at 4,000 rpm. In our analysis, the bigger heads could support a slightly longer camshaft with 4 to 5 degrees more duration. This would take a small amount of torque away from the 2,000- to 3,500-rpm band, but it would add torque and power from 4,000 rpm and up.

Dyno Results

The standard Twisted Wedge heads made 380 hp and 430 lb-ft of torque. The increased flow of the "R" heads made 400 hp and 450 lb-ft of torque—these numbers are comparable to the best big-cube engines of the '60s musclecar era!

Dyno Results of 351W with TFS "R" Heads

Engine rpm	Torque	HP
2,500	367	175
2,600	377	187
2,700	385	198
2,800	383	204
2,900	382	213
3,000	386	220
3,100	394	234
3,200	391	238
3,300	419	263
3,400	416	269
3,500	428	285
3,600	431	295
3,700	435	307
3,800	436	315
3,900	455	338
4,000	458	349
4,100	449	350
4,200	453	362
4,300	455	372
4,400	449	376
4,500	443	380
4,600	440	386
4,700	436	390
4,800	433	396
4,900	427	399
5,000	421	401
5,100	411	400
5,200	400	396
5,300	382	385
5,400	367	378
5,500	351	367

Dyno results prove the value of long rods and show that with the right combination of parts and calibration, factory EFI can match a carburetor for power. A heavy vehicle or one equipped with an automatic transmission requires a lot of low-rpm torque to provide neck-snapping acceleration. It is clear that our long-rod 351W is an ideal candidate for this. The torque curve is bountiful and fat, which means the engine will be tractable on the street.

THE 351 FORD SHOULD HAVE BUILT

When you consider that this was done using the Ford EEC-IV fuel-injection system with all its emissions strategies, 87-octane gasoline, a hydraulic roller cam that is but 10 degrees longer than a stock 5.0 cam, and that the engine is emissions-legal, the results are all the more impressive.

Summary

All in all, our long-rod SEFI 351W engine hit a grand slam, dispelling the notion that while fuel injection offers good cold start-up and driveability features, clean emissions, and respectable fuel mileage, it just can't keep up with a carburetor for power. We proved quite

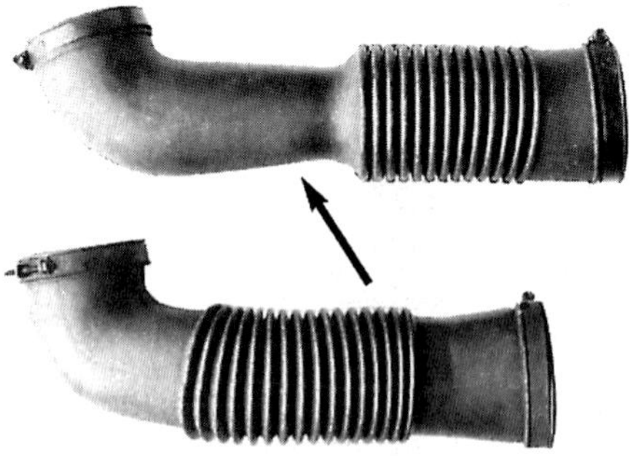

The wrong air-inlet tube between the mass air sensor and the throttle body can cost 15 to 20 hp. The top tube is the universal replacement tube for an '86 to '93 5.0 Mustang. Notice how it necks down (arrow) as compared to the original air tube (bottom). Use the original.

the opposite. Even with a good dual-plane intake and a streetable four-barrel carb, our 351W would have made about the same horsepower and

Effects of Restricted Inlet Air with Standard Heads

	Small Intake Tube		Large Intake Tube	
Engine rpm	Torque	HP	Torque	HP
2,500	378	180	379	180
2,600	384	190	386	191
2,700	401	206	400	205
2,800	416	222	419	225
2,900	412	228	414	230
3,000	409	234	409	234
3,100	406	240	406	240
3,200	406	247	405	246
3,300	422	265	422	265
3,400	427	276	427	276
3,500	426	284	426	285
3,600	422	289	421	289
3,700	419	295	419	294
3,800	422	305	420	303
3,900	425	316	425	315
4,000	427	325	427	325
4,100	428	334	428	335
4,200	431	344	431	344
4,300	428	350	430	352
4,400	423	354	425	358
4,500	415	355	422	361
4,600	410	358	418	366
4,700	403	361	414	371
4,800	396	362	409	374
4,900	390	364	405	378
5,000	384	365	400	380
5,100	376	365	389	378
5,200	365	362	376	373
5,300	354	357	363	366
5,400	342	351	349	358
5,500	322	339	328	345

These curves clearly demonstrate how sensitive the 351W engine is to intake breathing. The universal replacement air-inlet tube stocked by Ford cut the peak power by more than 15 hp, even with the small heads. Switching to an original '86 Mustang tube boosted power significantly and didn't hurt torque at all. With further improvements to intake breathing, the small-head version of our 351W could undoubtedly surpass 400 hp, though we imagine torque would max out at around 430 lb-ft.

In order to get the most airflow into the engine, we used a 75mm throttle-body unit from ACCUFAB. This unit is 15 mm larger than the stock 5.0 Mustang throttle body and yields a 33 percent increase in area. Although our 351W would have benefited from more intake airflow, a bigger throttle body would not have helped because the inlet opening in the Edelbrock manifold is 77mm (after Extrude-Honing).

Duttweiler says that fuel injectors work best when operating at no more than 85 percent of their maximum flow rating. At peak horsepower output, our 30 lb/hr Bosch injectors operated at more than 90 percent of their maximum flow rate. We strongly recommend 36-lb/hr units to reduce the duty cycle demanded at WOT and to improve the consistency of fuel delivery. Injectors larger than this will hurt low-rpm performance.

Fuel and Spark Delivery at WOT

Engine rpm	Ignition Timing (degrees)	Injector Pulse %
2,500	26.5	81.5
2,750	27.5	85.0
3,000	28.0	86.5
3,250	28.0	88.0
3,500	29.5	89.0
3,750	29.5	90.5
4,000	29.5	91.5
4,250	30.0	92.0
4,500	30.0	92.0
4,750	31.0	91.5
5,000	31.5	91.0
5,250	32.0	89.0
5,500	32.0	87.0
5,750	32.5	83.5

All other things being equal, the less timing an engine needs to make maximum power, the more efficient the combustion chambers are. Our engine only needed 32 degrees of total timing to produce maximum power. By comparison, most old-technology small-block Fords (and Chevys) need 35 to 40 degrees of total timing—again a measure of how efficient our long rods and optimized combustion chambers are.

would not have come close to the torque figure. *[Ed. note: Saueracker's long-rod Chevy 352, built for* Hot Rod *magazine, made 413 hp and 430 lb-ft of torque with a carb.]*

If you match the parts and assembly methods we used with our long-rod 351W, you can achieve the same power and torque figures. Since acceleration on the street is largely a function of torque, putting this anaconda under the hood of any car should result in the kinds of thrills and neck-snapping fun that only the musclecar era could deliver from a small-block—until now!

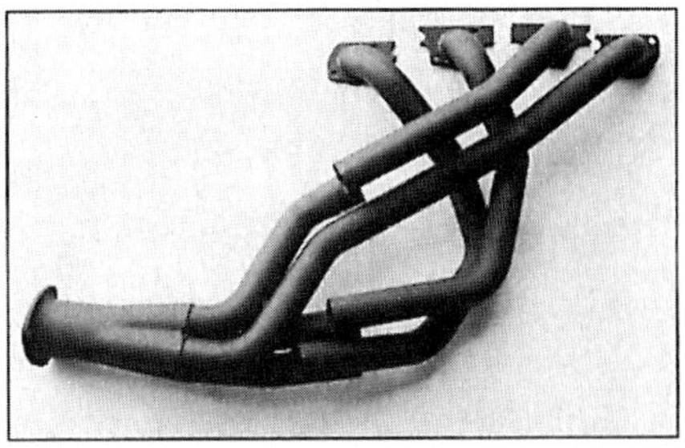

Hooker $1\frac{3}{4}$-inch-diameter primary headers (PN 6226) made massive torque down low. The individual primary pipes slip into the collector, making it easier to install the system once the engine is in the car.

Long-Block Parts

Part	Description	Source
Block:	Stock two-bolt 351W, bored 0.030-inch oversize	Local
Crank:	Stock 351W crank modified for long rods	Wayne's Engines
Rods:	Ford 400ci modified for 0.866-inch floating pins and 351W crank, with ARP rod bolts	Wayne's Engines
Bearings:	0.010-inch oversize main (PN MS-1432), 0.030-inch oversize rod bearings (PN CB-927)	AE Clevite
Pistons:	415-gram dished forging with 0.866-inch pin (PN 139105)	JE Pistons
Rings:	Pregapped 0.030-inch oversize moly (PN R-9902)	Speed-Pro (Dana)
Gaskets:	Overhaul kit with no- retorque head gasket (KS2329)	Fel-Pro
Cam and lifters:	Hydraulic roller cam, 220-degree duration (PN 444131), hydraulic retrofit roller lifters (PN 36532)	Crane Cams
Pushrods:	7.635-inch hardened steel for standard head (PN 11636), 7.335-inch hardened steel for "R" head (PN 27638)	Crane Cams
Rockers:	1.7:1 aluminum needle bearing rollers (PN 36757)	Crane Cams
Rocker studs:	7/16-inch 20 top, 7/16-inch 14 bottom screw-in (PN 99151)	Crane Cams
Oil pump:	High-volume (PN 224-102R)	Speed-Pro
Heads:	Twisted Wedge, 2.02-inch/1.60-inch valves (PN TFS-51400002), Twisted Wedge "R" head, 2.080-inch/1.60-inch valves (PN TFS-52400002) milled 0.070-inch	Summit Racing
Intake:	Extrude-Honed (290 cfm per runner) Edelbrock Performer (upper); 351W truck lower; two GGK phenolic spacers	Edelbrock

The entire long-block can be assembled for about $3,000 if you use the standard Summit TFS Twisted-Wedge head. The basic long-block price increases to about $3,800 if you opt for the "R" heads. Compared to what you have to pay for a carbureted crate engine, this is still pretty inexpensive, considering the power and torque the 351W can deliver with fuel injection in a fully emissions-compliant form.

SHOOTOUT SPECIAL

Editor's note: As a counterpoint to the money-is-no-object nature of the big-block shootout in this issue, we thought it would be a good idea to offer the reader an alternative—a shootout of engines built on a much more modest budget. Our idea was to provide a comparison of real-world, you-can-do-this-too small-block engines built to make serious power on pump gasoline. Well, we got that, and something else too—a major ego clash between HOT ROD staffers. To give you some behind-the-scenes flavor of how life has been in the office lately, we decided not to edit out all of the insults and character assassination, just those comments most offensive. Really, guys, leave your mothers out of this.—Drew Hardin

Hooker Headers supplied the 1¾-inch headers for this test. They are designed to be used in a late-model Mustang for 302 engines with TFS heads. The headers use an intermediate plate between the headers and the head. The plate required port-matching to prevent an exhaust restriction. It isn't easy making major horsepower on a budget, but it can be done. A shorter cam would reduce the horsepower but increase torque, which would give this engine a different personality on the street.

WINDSOR POWER

HOW TO BUILD A HIGH-HORSEPOWER 351W FORD ON A BUDGET!

By Jeff Smith

In the hierarchy of hot rodding, Fords are the Rodney Dangerfield engines. They get no respect. This time was no exception. HOT ROD staffer Cole Quinnell proposed a small-block horsepower shootout between Chrysler, Chevy and Ford. But this one had a twist. The budget for the buildup was $2400. Cole chose the Chrysler, Rob Kinnan took the easy way out with the small-block Chevy, and I took the Ford even though I thought it would be handicapped due to the higher cost of hi-po Ford parts.

Unfortunately, this shootout became a two-way contest when Quinnell discovered the Mopar couldn't run with the big dogs at a budget of $2400. Sure, the Chrysler won the big-block shootout (thanks to a no-limit budget and an admirable effort by Herb McCandless and crew), but there just aren't any parts to build a killer A-motor that you could call "affordable." And as it turned out, none of us could keep to that original budget.

The goal of this shootout was to achieve peak horsepower on a budget using pump gasoline. This meant big heads, a long-duration cam and lots of rpm. Sadly, this results in a couple of engines on the ragged edge of "streetable," unless you actually like an engine with single-digit idle manifold vacuum and soggy torque below 4000 rpm. We expected peak horsepower to happen at around 6500 rpm. The problem is that Uncle Henry never intended stock rods and cast cranks to run at 6500 rpm and higher. You can get away with it for awhile, but not forever.

HEADS UP

After discussing the 351W's potential with Ford guru Chris Kaufmann, we chose the iron version of the TFS head. Not only does it produce outstanding intake and exhaust flow numbers right out of the box, but a pair of these bare heads is available for $695 from Summit or TFS. We squeezed the budget until it screamed for mercy and managed to assemble the heads for well under the TFS $1195 list price for a pair of assembled iron heads. However, we scrimped on a few items, so unless you really know what you're doing, we'd suggest buying the heads complete from TFS to save time and aggravation.

Thinking that Kinnan and his band of cutthroat Chevy guys might be tempted to cheat like the dickens, we decided to be equally devious. While the TFS heads are killers right out of the box, Kaufmann steered us to Charlie Hupp at Portmasters to give us some guidance on porting these heads. Since there wasn't room in the budget for a professional port job, I applied the basics detailed in two previous HOT ROD porting stories ("Port-A-Ble Power," May '91 and "Chamber Music," September '91).

CAM IT UP

To take advantage of all this airflow, a longer-duration cam is necessary to keep the valves open long enough at high rpm to feed those hungry cylinders. We used the Dynomation computer engine simulation program to help us choose a camshaft. After some experimentation, we decided on a Comp Cams 306 solid-lifter, single-pattern camshaft with .590-inch valve lift and 260 degrees of duration at .050-inch tappet lift. The program also indicated peak horsepower at 6500. This turned out to be right on the money, even though the program estimated more torque and horsepower than we actually made.

This was important information. We didn't want to spin this engine more than 6500 rpm, since our valvesprings were limited to 105 pounds of seat pressure and around 330 pounds at maximum lift. As it turned out, the gentle Comp Cams profile kept the engine out of valve float. The rest of the valvetrain consisted of a

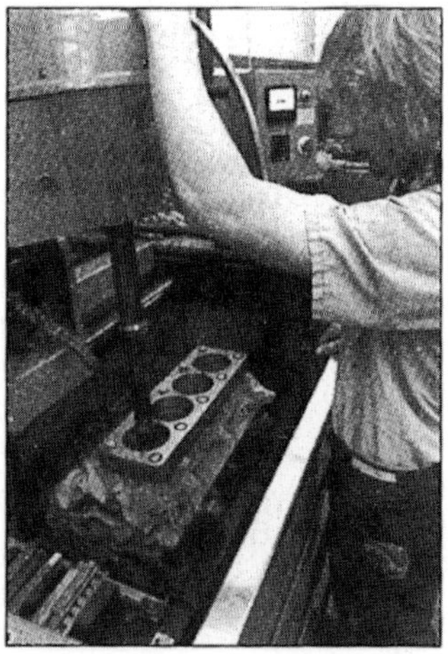

Jim Grubbs Motorsports performed all the machine work on our 351W. Here, Todd Roselund torque-plate hones the block. JGM also decked the block and rebuilt the rods.

Bill Mitchell's trick rod vise was very useful in torquing the rod bolts in order to check bearing clearances.

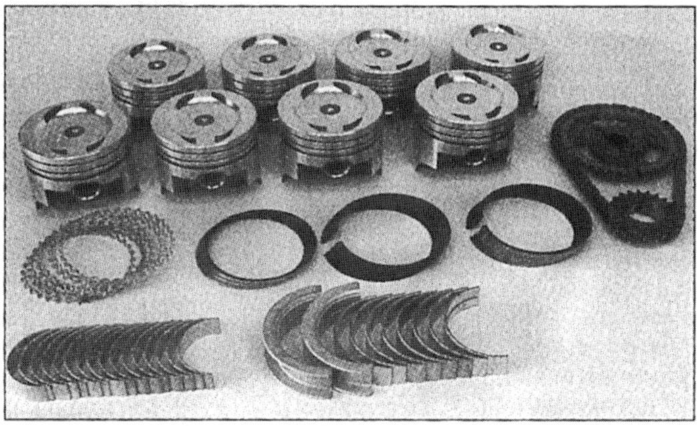

We used TRW replacement forged, dished pistons along with a set of 5/64-inch single moly Sealed Power rings, .010-under bearings, and a stock timing chain gear and set. The dished pistons were a good choice since the TFS heads offer a small 60cc chamber. This kept the compression to a realistic 9:1.

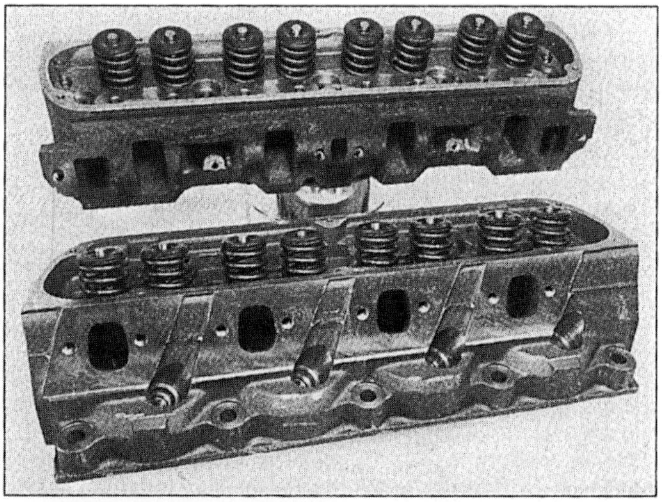

Mocking up the engine ahead of time revealed that the intake valve hit the piston. So fly-cutting the pistons with wider valve reliefs was necessary to generate clearance. The minimum valve-to-piston clearances are .100 inch for the exhaust and .080 inch for the intake.

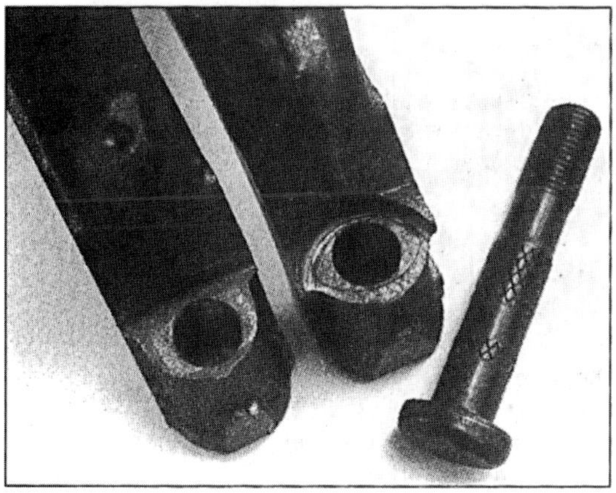

Early 351W rods cut across the rod for the bolt head (*left*). Later truck rods (*right*) use a circular cutout that offers more strength. We also used ARP's 190,000-psi bolts to keep everything together at 6500 rpm.

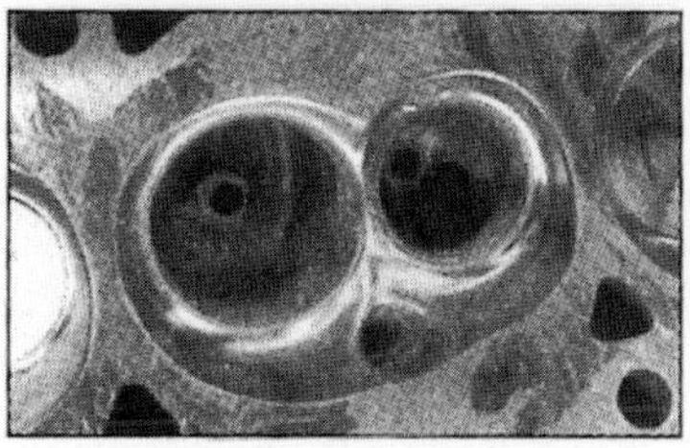

The TFS heads are available two ways. The bare heads come with a valve job and honed guides for $695. You can also obtain them for $1195 complete except for studs and guideplates. The iron TFS heads weigh a ton, but they kick ass on airflow, which is the only way to make horsepower.

set of Harland Sharp roller rockers, steel retainers and machined keepers.

SHORT STUFF

Compared to the tricked-out heads and cam, the short-block is painfully stock. We used stock replacement, fully dished TRW forged pistons which, combined with the modified chamber heads and .005-inch deck, create a 9.1:1 compression ratio. The pistons use single moly, 5/64-inch Sealed Power rings, stock rods with ARP bolts, and a stock oil pump and timing chain set also from Sealed Power. The budget wouldn't even allow a windage tray in the oil pan! Thankfully, we found room for Fel-Pro gaskets, which we used throughout the engine.

MAJOR FLOW

To determine the proper camshaft, we flowed the TFS heads just as they came out of the box on Extrude Hone's SuperFlow 600 flow bench. Then we ported the heads based on previous HOT ROD how-to stories. Charlie Hupp of Portmasters was also helpful in directing work on these heads.

The two flow sheets reveal both the out-of-the-box flow numbers (which are outstanding) and the flow after porting. All tests were performed with the stock 2.02/1.60-inch valve sizes at a test depression of 28 inches of water.

	TEST 1		TEST 2			
	TFS Stock Heads		TFS Ported Heads		Increase	
Valve Lift	Intake (cfm)	Exhaust (cfm)	Intake (cfm)	Exhaust (cfm)	Intake (cfm)	Exhaust (cfm)
.150	97	96	102	84	+ 5	+12
.250	158	132	173	132	+15	---
.350	203	159	225	164	+22	+ 5
.450	233	175	261	182	+28	+ 7
.550	250	184	278	193	+28	+ 9
.600	259	187	285	195	+26	+ 8
.650	259	190	293	199	+34	+ 9

PARTS, NUMBERS & PRICES

COMPONENT	PART NO.	PRICE	SOURCE
Pistons, TRW, 17cc dish	L2378	$200	PAW
Rod bolts, ARP	150-6004	39	CP
Camshaft, Comp Cams 306S	31-336-4	114	PAW
Lifters, mechanical	---	40	PAW
Intake, Edelbrock Victor Jr.	2981	189	KPI
Rebuild kit, deluxe (rings, bearings, gaskets, pump, timing set)	---	142	PAW
Freeze plug kit	P-108	3	CP
Oil pump driveshaft	8153	14	CP
Cylinder heads, TFS, iron, bare	---	695	TFS
Valves, 2.02/1.60 stainless	8440	75	CP
Springs	2926	45	CP
Retainers	3437	30	CP
Keepers	4133	25	CP
Studs, ARP	100-7101	25	CP
Guideplates	WLB-314410	24	TFS
Pushrods, 5/16 .080 wall	1730	90	CP
Rocker arms, Harland Sharp	HS4003	178	CP
Subtotal		$1928.00	
MACHINE WORK			
Hot tank and mag		$ 64	
Balance crank and rods		150	
Rebuild rods		96	
Bore and hone		152	
Deck block		80	
Fly-cut pistons		120	
Deck heads and manifold		100	
Subtotal		$ 762.00	
TOTAL		$2690.00	

KEY TO SOURCES
- CP Competition Products
- KPI Kaufmann Products, Inc.
- PAW Performance Automotive Warehouse
- TFS Trick Flow Specialties

The valvetrain required quite a bit of price shopping to stay close to the budget. Steve Mugerauer's Competition Products in Oshkosh, Wisconsin, put us in great shape by supplying almost everything for the valvetrain including the Harland Sharp rockers, valvesprings, retainers, machined keepers, freeze plug kit and even a Wolverine high-performance oil pump driveshaft. Because of the wider-than-stock valve spacing, the guideplates must come from TFS. The cam is a Comp Cams solid lifter 305 with 260 degrees of duration at .050-inch tappet lift.

With the stock Ford 1.6:1 rocker ratio, gross valve lift is a healthy .590 inch. This required an honest 1.800-inch installed height, but our first attempt gave us only 1.760 inch! This put the springs virtually into coil bind at max lift. Competition Products supplied us with a set of Crane machined keepers that added the extra .050-inch clearance we needed while also keeping the retainer away from the valve-guide seal. Minimum clearance for these two specs is .050 inch. We were very close to the minimum, so if you are considering a similar engine, make sure to check this carefully.

Our search for a less expensive Edelbrock Victor Jr. intake took us only as far as Kaufmann Products, which had the best price for this excellent intake manifold. We port-matched the intake to the TFS heads and found that a 1-inch open plenum spacer was the plan for peak power.

Ford uses a thrust plate to keep the cam in place and requires these two thin-head bolts to retain the plate. Normal bolt heads will hit the cam gear. We also degreed the cam to ensure it was installed properly at Comp's 106-degree intake centerline.

One of our secret weapons in this engine buildup was the help of our pals at Jim Grubbs Motorsports (JGM), who did all the machine work on our Ford. They also helped us mock up the engine at their shop to check for valve-to-piston clearance and deck height.

Ken Duttweiler offered the use of his dyno to pretest the Ford to make sure we squeezed the last ounce of power out of this 351W. Timing, jetting and valve lash added more than 30 horsepower over our initial power pull.

Checking valve-to-piston clearance is a necessity with the TFS heads since the valves have been moved apart to allow them to be bigger. In our case, the intake valve hit the piston in mock-up, which required fly-cutting the pistons. JGM's excellent machine work made assembling the 351W much easier and certainly contributed to the 351W's strong power. There's no substitute for quality machine work.

TAKING NO CHANCES

Since we were ahead of the game, Ken Duttweiler was gracious enough to allow us one day on his dyno to pretest the engine. Our goal was 450 horsepower, but just like those soap operas on TV, we're going to leave you hanging about the results. Let's just say we came mighty close. Of the things we didn't try but planned to do on the final test was a slight change of cam timing and a test of a load of synthetic 30W racing oil from Red Line. Since the engine was brand-new, we broke it in on straight 30W engine oil.

Just for fun, we plugged our power curve into Patrick Hale's Quarter computer simulation program and came up with a 3400-pound Mustang that, with a 4.10 gear, 28-inch-tall slicks and a well-tuned chassis, could run very low 12s without too much trouble. Of course, if you're willing to run race gas, this combination with 11:1 compression could make a little more than 450 horsepower! That's decent power in anybody's book, especially on a budget. **HR**

SOURCES

Automotive Racing Products (ARP)
Dept. HR07
250 Quail Ct.
Santa Paula, CA 93060
805/525-5152
800/826-3045

Competition Cams
Dept. HR07
3406 Democrat Rd.
Memphis, TN 38118
901/795-2400
800/999-0853 (Cam Help)

Competition Products
Dept. HR07
3200 Medalist Dr.
Oshkosh, WI 54901
414/233-2023

Ken Duttweiler Performance
Dept. HR07
1563 Los Angeles Ave.
Saticoy, CA 93004
805/659-3648

Extrude Hone
Dept. HR07
8800 E. Somerset Blvd.
Paramount, CA 90723
310/531-2976

Federal-Mogul Corporation
Dept. HR07
P.O. Box 1966
Detroit, MI 48235
810/354-7700

Jim Grubbs Motorsports
Dept. HR07
28130 Avenue Crocker
Unit 331
Valencia, CA 91355
805/257-0101

Hooker Headers
Dept. HR07
1024 W. Brooks
Ontario, CA 91762
909/983-5871

Kaufmann Products
Dept. HR07
12420 Benedict Ave.
Downey, CA 90242
310/803-5531

Performance Automotive Wholesale (PAW)
Dept. HR07
8966 Mason Ave.
Chatsworth, CA 91311
818/998-6000 (orders and tech line)
818/407-2650 (customer service)

Portmasters
Dept. HR07
545 W. Allen Ave. #2
San Dimas, CA 91773
909/394-5929

Summit Racing Equipment
Dept. HR07
P.O. Box 909
Akron, OH 44398-6177
216/630-0200 (orders)
216/630-0230 (customer service)
216/630-0240 (tech line)

Trick Flow Specialties (TFS)
Dept. HR07
1248 Southeast Ave.
Tallmadge, OH 44278
216/630-1555

VP Engineering, Inc. (Dynomation)
Dept. HR07
2921 Patricia Dr.
Des Moines, IA 50322
515/ 276-0701

WINDSOR WONDERLAND

ENLARGING FORD'S 351

By Steve Anderson

Photo: Lynne McCready-Coombes

It's gone on for more than 70 years, and the scars run deep as the battle between the Bow-Ties and Blue Ovals continues to this day. Chevy and Ford owners defend their hallowed territory to the last cubic inch, and it looks as if they will continue this battle ad nauseam.

While Chevy owners have won their fair share of encounters, Ford owners are making better and better Mouse traps. Two such devices were recently created by Andrew Pozzi and George Caven of Industrial Machine in Santa Rosa, California, and Roger Decker, owner of Canton Automotive Machine Service in Canton, Ohio. Based on the 351 Windsor, both companies have joined an unusual combination of hardware and machining processes to come up with new configurations of a familiar small-block Ford. While they took different routes to obtain their objectives, both achieved the intended goal of putting the Bow-Ties square in their sights.

397 WINDSOR

For Pozzi and Caven, the idea of a 397 Windsor came after they considered the involved process of replacing the Windsor with an all-too-large 460 Ford. This would have required new mounting hardware and plumbing and considerably more time and money than either of them could afford. After examining the various options from Ford, along with numerous performance outlets, they achieved their goal through typical hot rodder budget tactics: They built their own stroker engine. **HR**

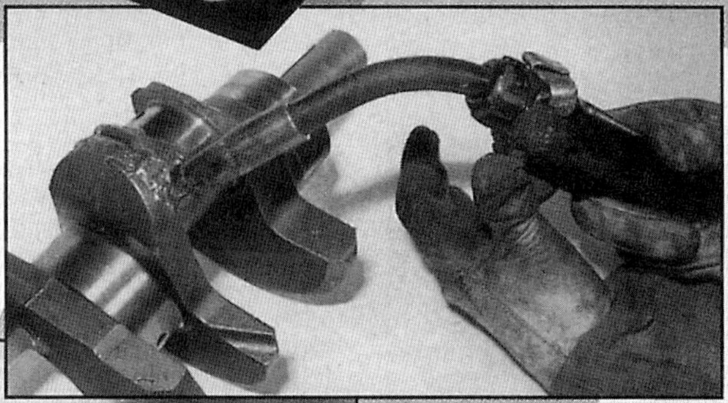

To increase the stroke of the stock 351 crankshaft, the outside of the rod journals was enlarged by .400 of an inch by a crankshaft specialist. This was done with great care because a weld with poor penetration and/or air pockets will eventually result in journal fatigue and failure. As the area was built up across the upper bearing surface, a wire brush was used to clean off any slag that would impede the welding process.

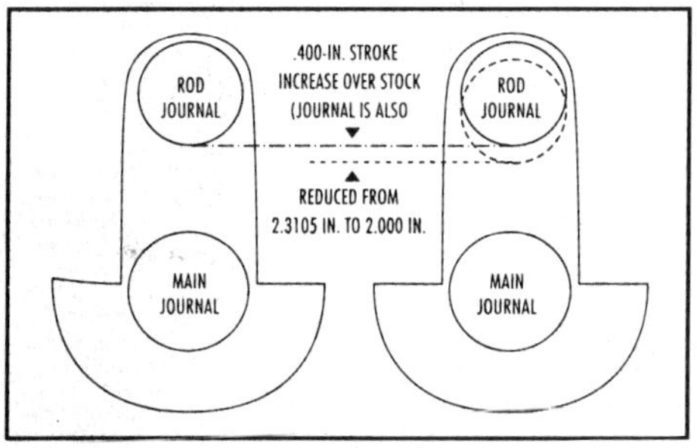

Once enough thickness was welded to the outside of the crank, the journals were remachined, using a crank grinder to increase the stroke of the crankshaft by .195 of an inch. The journals were also reduced in diameter during the machining process from 2.3105 to 2.00 inches to accept the smaller-journal Buick rods.

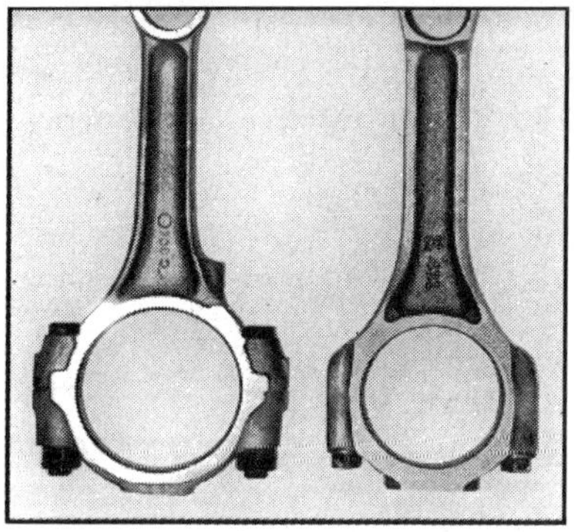

A better rod-length-to-stroke relationship was gained through the use of 431 casting number Buick 231 V6 rods (right), which are .011 of an inch longer than the stock 351 Windsor rods. This view also shows the smaller diameter of the Buick's rod journal.

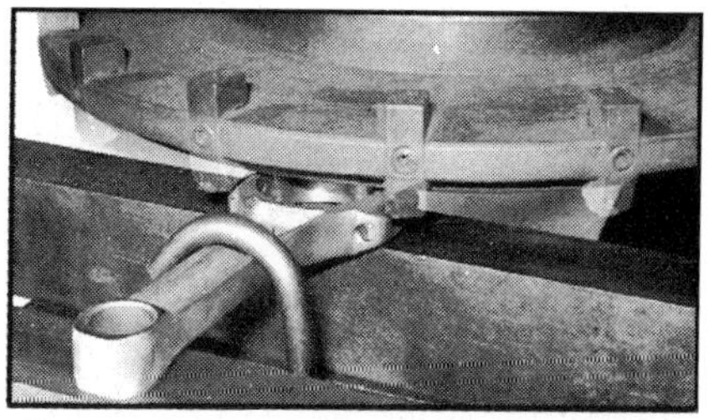

For the Buick rod to fit into the Ford crankshaft rod journal, .012 of an inch was milled off the offset side of the connecting rod to fit the Ford's .835-inch bearing surface width. A set of Sealed Power CB610P rod bearings completes the package.

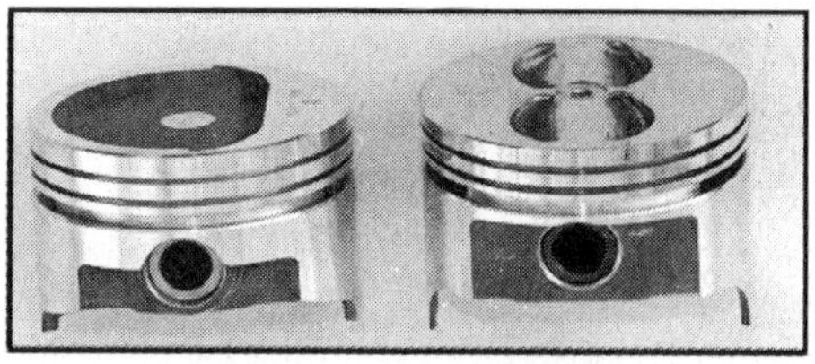

While this 397 uses 4.030-inch Speed-Pro 7058PA turbo pistons (left), more power can be gained by adding a set of higher compression Speed-Pro 7064P flattop pistons with four valve reliefs (right). The final static compression ratio will be determined by deck height, gasket thickness, and combustion volume.

SOURCES

Canton Automotive Machine Service (C.A.M.S.)
Dept. HR
3048 Navarre Rd. S.W.
Canton, OH 44706
(216) 453-8383

Industrial Machine
Dept. HR
928 N. Dutton Ave.
Santa Rosa, CA 95401
(707) 542-8956

Andrew Pozzi
Dept. HR
850 Second St. Ste. B
Santa Rosa, CA 95401
(707) 545-5874

Speed-Pro/Sealed Power Corporation
Dept. HR
100 Terrace Plz.
Muskegon, MI 49443
(616) 724-5200

Trick Flow Specialties (TFS)
Dept. HR
6077 Back Orrville Rd.
Wooster, OH 44691
(216) 682-8331

Will-Burt Company
Dept. HR
169 S. Main St.
Orrville, OH 44667
(216) 682-7015

Thanks to John R. Smith for his extensive research and writings on Roger Decker's 416 Windsor.

416 WINDSOR

Unlike the 397 Ford, Roger Decker's Canton Automotive-built 416 Windsor utilizes a 4.030-inch bore and an offset-ground 400M crankshaft with a 4.075-inch stroke to create this thumper Ford.

The Ford 400M crank (center) is the recipient of major modifications that include reducing and offset grinding the rod journals to increase the stroke by .075 of an inch, incorporating a number of machining operations to the crank snout, tail "slinger," and trimming the counterweights. Heavier Mallory metal is required for proper balancing.

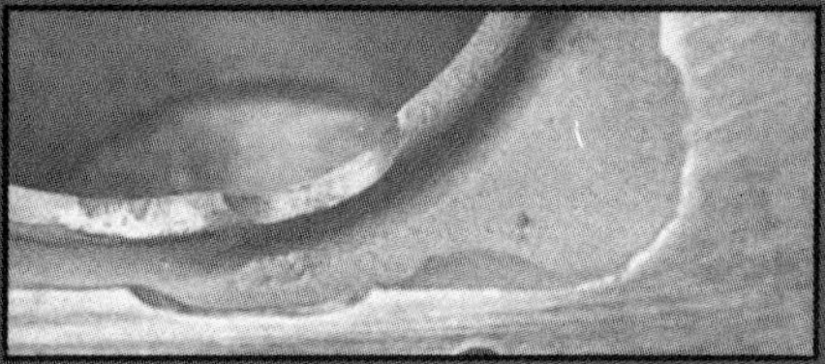

To provide adequate clearance for the 400M crank, the oil-pan rail and cylinder-wall base must be notched. This must be done properly to prevent block damage or contact between the rod and block.

Another unusual component found in the 416 Windsor is a Chrysler 360 rod that was narrowed at the crankshaft end .050 of an inch (.025 of an inch per side) for proper clearance.

While the bottom side of the pistons and the crankshaft counterweights have been radiused, there is only .060 of an inch of clearance. Measurements must be accurate!

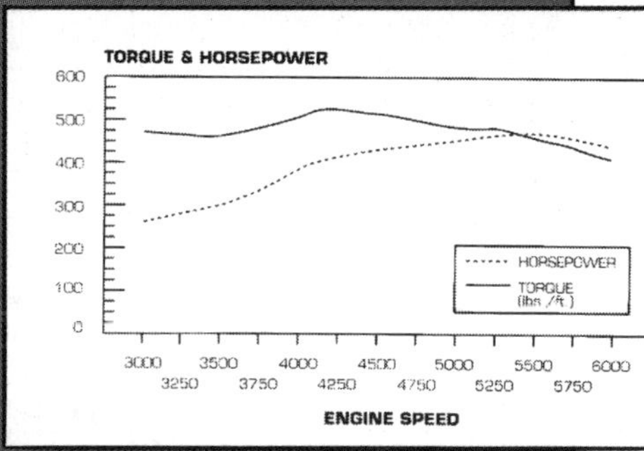

Fitted with an Edelbrock Victor Jr. intake manifold and a Holley 850-cfm 4-barrel with No. 76 jets, Decker's 416 Windsor motor produced 514 lbs.-ft. of torque at 4250 rpm and 485.9 hp at 5750 rpm.

The Trick Flow Specialties/Will-Burt Street Heat aluminum cylinder heads used on the 416 were heavily massaged for maximum flow and fitted with 2.055 intake valves and 1.600 exhaust valves. Crower valvetrain components keep it all together.

HORSE POWER

By Al Kirschenbaum

Have you ever noticed that the more some things change, the more they seem to remain the same? Take Ford's Mustang engine line-up, for example. When the Dearborn autoworks presented their little prototype pony (then called the Mustang I) in the early Sixties, it was powered by a foreign-built 4-cylinder Ford powerplant. Now, nearly two decades later, the freshest factory Mustangs feature 4-bangers as standard, and their one-time 260-inch, then 289, then 302, then 255-cube base eight has been opened back up to 302 muscular cubic inches.

What *has* changed forever, however, is nearly everything powerful in between. In the 20-year period, the Ford ponycar line has seen all manner of motors—from muscular to minuscule—and nearly all have made their own scorched rubber marks on modern motoring.

For the purposes of this top-option review we're condensing history and looking only at the line's highlights. Granted there are a few notable pony motors that we're flat-out ignoring (the 390 and 427-inchers in particular and all Shelby engines in general), but the subject of an era's supermotors is entirely too broad for *the* final word on all the ultimate Fords. So for the moment, at least, we'll touch only on the highest of the Mustang's mechanical high points and look to future issues of HOT ROD for all of Ford's old-but-better ideas. **HR**

289 HIGH PERFORMANCE

The Mustang's very first taste of high-output breeding came late in 1964 when Ford fitted their popular ponycar with a solid-lifter model of their 289-cube small-block V8. The little engine featured longer camshaft timing, free-breathing exhaust manifolds, a high-strength crankshaft and connecting rods, threaded rocker arm studs, dual-point ignition, and a low-restriction air cleaner. Check out our spec chart for the HiPo motor's internal details and see HOT ROD, December 1980 for additional information on the complete Mustang/motor combination.

BOSS 302

In many ways, Ford's Boss 302 engine was a hot rodder's fantasy executed in iron. But far from being an all-too-typical home-brewed exercise in excess, the factory package was built on time-tested hardware with a solid engineering edge. In other words, what the motor folks in Dearborn did was simply to refine—hot rod style—their already proven 289.

A high-strength "stroker" crank bumpe the cubes to 302 while an extra pair of b at each center main added structural support to the forged steel shaft. New cylinder heads, Cleveland-style, were als included, and their arc-shaped combusti chamber roofs permitted the fitting of extra-large valves which, in turn, fed through even larger ports and passages.

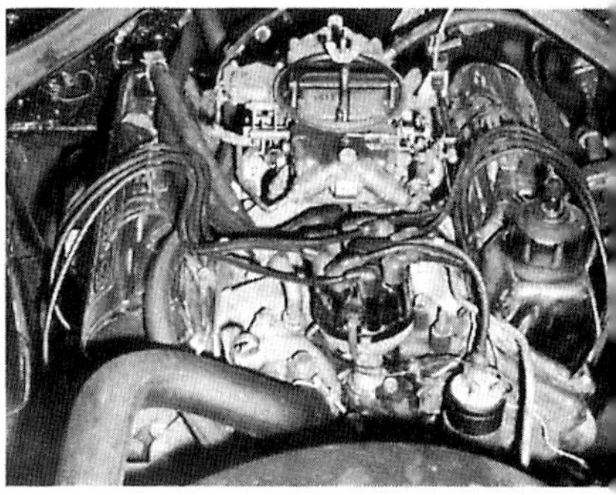

MUSTANG MUSCLE MOTORS

							INTAKE SYSTEM								
							MANIFOLD		CARBURETOR		CAMSHAFT			VALVE SIZES	
ENGINE	YEAR(S)	HORSE-POWER	TORQUE	COMPRESSION RATIO	BORE	STROKE	TYPE	MATERIAL	MFG'R	FLOW RATING	DURATION	LIFT	TYPE	INTAKE	EXHAUST
289 High Performance	1964-1968	271@ 6000 rpm	312@ 3400 rpm	10.5:1	4.00	2.87	1x4 180° dual-plane	cast-iron	Autolite	480 cfm	310°/310°	.460	mechanical	1.78	1.45
Boss 302	1969-1970	290@ 5800 rpm	290@ 4300 rpm	10.5:1	4.00	3.00	1x4 180° dual-plane	aluminum	Holley	780 cfm	290°/290°	.477	mechanical	2.23 (1969) 2.19 (1970)	1.71
Boss 351	1971	330@ 5400 rpm	370@ 4000 rpm	11.7:1	4.00	3.50	1x4 180° dual-plane	aluminum	Autolite	(1)	290°/290°	.477	mechanical	2.19	1.71
428 Cobra Jet	1968-1970	'68: 335 @ 5600 '69-'70:335 @ 5200	'68: 445 @ 3400 '69-'70:440 @ 3400	10.7:1	4.13	3.98	1x4 180° dual-plane	cast-iron	Holley	735 cfm	270°/290°	.481/.490	hydraulic	2.09	1.66
Boss 429	1969-1970	375 @ 5200 rpm	No data	10.5:1	4.36	3.59	1x4 180° dual-plane	aluminum (low-profile)	Holley	735 cfm	'69 (all S, early T) 288°/292° '69-'70 (later T) 300°/300°	.493/.501 .509	hydraulic (3) mechanical	2.28	1.90

e '69 Boss intake valves actually proved
e too large for suitable street response,
' they were later downsized for the
owing model year.) Free-breathing was
her promoted through a generously
oortioned inlet manifold and a
n-volume Holley carb. And on top of
the entire intake works could be even
her enhanced through the use of the
ory's optional over-hood Ram Air.
ails on the Boss motor's internals are
uded in the Muscle Motor chart.

SS 351

ike the other Boss Mustang models, the
351 wasn't built to satisfy a race
ctioning body's minimum production
nands, but rather to suit a variety of

w age" marketing conditions. "Tamer"
robably the best word to describe the
0 or so Boss factory fastbacks that rolled
during the single year's run, and all
e designed for smoother, more
etable response.
he cars' performance level was
ntained through the use of the same
nshaft specs and the same style cylinder
heads as the Boss 302 along with similarly engineered internals. Ram Air was again installed while oiling and cooling improvements (including the power steering fluid cooler seen just forward of the driver's side valve cover) helped the relatively short-lived "middle" Boss to close out Ford's muscle generation in very street-wise style.

428 COBRA JET

Civilized muscles sprouted under Mustang bonnets when Ford upgraded their 428-cube Police Interceptor FE-series engines around 427 race-style cylinder heads in late '68. The resulting Cobra Jet powerplants packed more torque than the Mustang's street rubber could handle, but a lack of initial traction was hardly enough to keep a good option down. In fact, the 'Stangs that were so equipped became one of the most popular street machine assemblies ever offered.

The big wedge fed on fresh air pulled in from above the hood through a "Shaker" scoop/air cleaner that was optional on the

'69 and '70 Mach I models. There's little doubt that this neat Ram Air arrangement was what helped the CJ package to put Ford right up there again with the street fleet's big-inchers from Chrysler and GM.

Details on this Mustang's under-hood mechanicals can be found in the chart, while additional data on the 428CJ is provided in HOT ROD, December 1979.

BOSS 429

Of the five engines reviewed here, only the Boss 429 was really a new piece for the street rather than an outgrowth of another motor. At least it didn't **look** like any V8 that came before it. Ford decided to shoehorn 852 of these Blue Crescent-based engines into '69 Mustangs (505 were assembled for 1970) when a minimum of 500 production models equipped with the engine were required in order to qualify it for competition in NASCAR's Grand National stocker ranks. Fortunately for enthusiasts, the company decided to have their Kar Kraft subsidiary fit the staggered-valve monster motors to the sportier Mustang models rather than to the full-size Galaxies as they'd initially intended.

Equipped with exotic aluminum cylinder heads that were really considered the life-breath, heart, and soul of the "Boss Nine" package, the few 429 Mustangs that did find their way streetside were actually down on both bottom-end torque and top-end carburetion. Huge intake ports and extra generous oval-shaped exhausts made the Mustang better-suited to the high-banks than to the highways, but as the rules would have it, only Ford's larger cars ever saw big-time roundy-round competition with the Boss 429 aboard. **HR**

		CRANKSHAFT					
R			MAIN BEARING JOURNAL	CONNECTING RODS			
	IGNITION SYSTEM	MATERIAL	DIAMETER	MATERIAL	LENGTH	JOURNAL DIAMETER	FRESH AIR INTAKE SYSTEM
	dual-point w/o vacuum advance	high-nodular cast iron	2.2486	forged steel	5.155	2.1232	not available
	dual-point w/ dual vacuum advance (2)	forged steel (4)	2.2486	forged steel	5.150	2.1232	optional "Shaker" hood w/scoop attached to air cleaner
	dual-point w/ dual vacuum advance	high-nodular cast-iron (specially selected)	2.749	forged steel	5.78	2.311	optional ducting between dual hood scoops and air cleaner
	dual-point w/ dual vacuum advance	high-nodular cast-iron	2.749	forged steel	6.488	2.438	hood scoop on late 1968 models, optional "Shaker" hood in 1969-'70
st	dual point w/ dual vacuum advance (2)	forged steel	3.00	forged steel (3)	6.549 ('69 S) 6.605 (late '69-'70 T)	2.50	manually controlled hood scoop ducted to air cleaner

Notes:

(1) Ford's Autolite carburetors were not rated by official flow figures. Estimates for the Boss 351's model 4300-D spread-bore carb fall into the 600-750 cfm range.

(2) Both the Boss 302 and the Boss 429 Mustangs built for 1970 were equipped with rev limiters. These units were individually calibrated for each engine and were designed to limit and then cut voltage to the coil at predetermined points. Fortunately, Ford retained the standard under-hood wiring harness even when a limiter was fitted, and all that was required for "unlimited" off-road operation was some minor plug-in rewiring.

(3) Two street versions of the Boss 429 were built. The first few hundred were designed for heavy-duty use and were designated "820-S." They use a super tough crank, rod, and piston combo. Later "street" versions are called "820-T" and use a different crank, rod, and piston. All "S" engines and early "T" engines came with a hydraulic cam. Late '69 "T" engines and all '70 "T" motors use a mechanical cam, the same one as used in 429-SCJ engines. There was also a race-only NASCAR 429 with different heads, rocker arms, and reciprocating assembly.

(4) The Boss 302's steel crankshaft was forged with hollow throws in 1969 while the 1970 versions were solid shafts which, unlike the earlier models, weren't cross-drilled for oil flow.

AN AFFORDABLE IRON FORD 302/351W CYLINDER HEAD!

Windsor Gets a Head

By Marlan Davis

AN EXCLUSIVE HOT ROD STORY

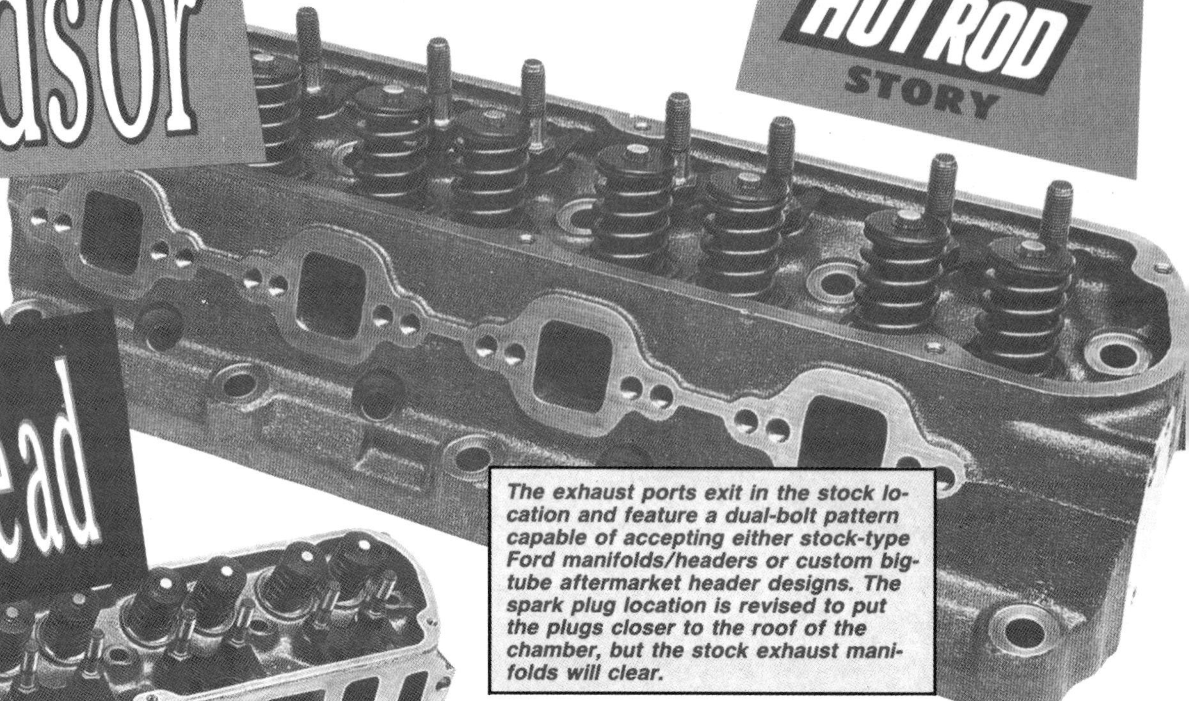

The exhaust ports exit in the stock location and feature a dual-bolt pattern capable of accepting either stock-type Ford manifolds/headers or custom big-tube aftermarket header designs. The spark plug location is revised to put the plugs closer to the roof of the chamber, but the stock exhaust manifolds will clear.

The intake port entry remains in the stock location, but the overall contour is slightly enlarged to better match existing Ford intake gaskets.

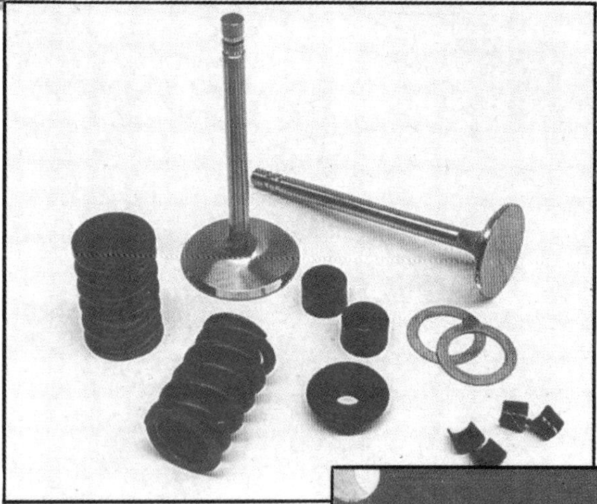

World Products machines the heads to accept Ford 289/302 rocker arms and small-block Chevy-style pushrod guideplates, valves, valvesprings, and retainers along with nylon or umbrella-style valve seals. Chevy 2.02/1.60-inch valves are standard.

The heart-shaped combustion chamber is all the rage in today's high-tech wedge heads. The 64cc chamber utilizes Chevy "fuelie" valves and unleaded-fuel-compatible steel-insert exhaust seats.

World Products, the company that gave the world the first cast-iron, high-performance, aftermarket small-block Chevy heads, is now in the process of breathing new life into Ford's Windsor dynasty. For years, proponents of wedge-style (inline valve) small-block Ford heads have played second fiddle to the canted-valve Cleveland boys. They took for granted that there was "just no way" you could get wedge heads on a Ford small-block to keep up with the Chevys. True, the last 10 years have brought us several aluminum wedge-style heads that represent steps in the right direction, but their relatively high prices and (in the case of at least one of the offerings) nonstock exhaust-port exit locations made them uncompetitive against Chevy-style designs.

This is where the Dart II World Products heads come in. When the company decided to try its hand at making a Ford head, it approached the task by applying proven cylinder head design technology. Internal-combustion engines all operate according to the same principles of thermodynamics, and they don't care whose name's stamped on the valve cover! Over time, Chevy wedge intake and exhaust ports had proven superior to Ford ports; state-of-the-art wedge chambers (regardless of manufacturer) gravitated away from "bathtub" shapes toward the highly efficient heart-shaped designs; and aftermarket valvetrain-component parts availability and pricing were primarily geared to the Chevy motor.

World started from scratch, modifying the Chevy port design to fit into the Ford envelope. It kept the original Ford intake-port-entry and exhaust-port-exit locations to allow retention of the existing intakes and headers. However, beyond the intake opening/exhaust exit, the ports were dramatically recontoured to optimize flow, with higher roofs and a more concentric short-turn-radius transition to the valve bowl. The intake runner volume displaces 195 cc, about the same as Dart's Chevy head. On the exhaust side, the valve seat location was moved .040 inch, to allow room for larger 2.02-inch intake/1.60-inch exhaust Chevy valves. Because Chevy valves are .050-inch longer than the Ford O.E.M. design, the new head's valve seats are located .060-inch deeper inside the revised, heart-shaped combustion chamber to permit use of existing adjustable-type Ford rocker arms and pushrods. To enhance combustion of the air/fuel mixture, the 14mm taper-seat spark plug hole location is closer to the chamber roof and exhaust side of the head. While the high-swirl chamber configuration represents a departure from previous production Ford designs, it's said to be compatible with all Ford piston-dome shapes. At 64 cc, chamber displacement is a midrange compromise among the numerous Windsor chamber sizes used through the years; but the beefy, .400-inch-thick deck can easily be milled to produce significantly smaller chamber volumes, as desired by the customer.

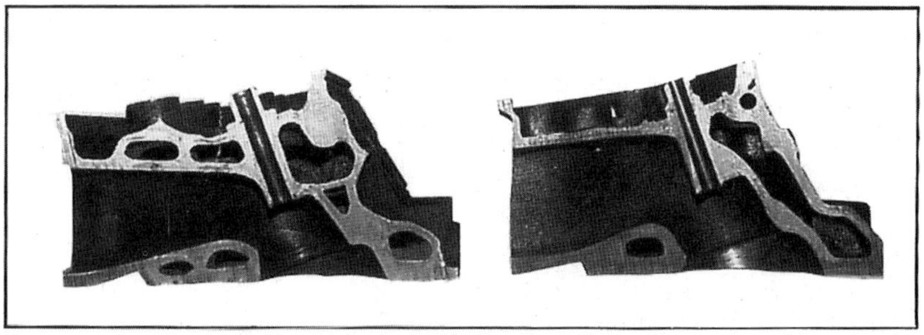

Compared with stock Ford heads (left), the Dart II head's port roof is raised considerably beyond the intake entrance, and the contours of its short-turn radius are rounded off.

Fords traditionally have weak exhaust ports (stock head, left). Careful massaging (right) has enabled Dart engineers to achieve exhaust-flow numbers comparable to Chevy heads.

The relocated valvetrain allows for an adjustable valvetrain assembly. Ford Motorsport off-the-shelf, small-block Chevy, or (for heavy-duty use) big-block Chevy adjustable rocker arm studs screw right into the heads, interfacing with small-block Chevy pushrod guideplates, Chevy 1.25-inch o.d. valvesprings, Ford 289/302 rocker arms (with small-block Chevy or Ford rocker studs), and Ford pushrods to complete the valvetrain (see parts list).

The exhaust face of the heads is dual machined for both the standard Ford exhaust bolt pattern as well as a custom spread pattern for greater clearance of custom big-tube header flanges. Half-inch-diameter head-bolt holes allow the heads to fit on all Ford Windsor-style small-blocks, including early 351W blocks. Later blocks with 7/16-inch head-bolt holes must use a special Ford stepped head bolt. Finally, the heads are delivered with the Ford Thermactor (smog pump) passage drilled and tapped lengthwise through the head. However, the corresponding intersecting passages running from the main passage to each exhaust port are not drilled all the way through into the ports. Instead, a mark (dimple) is cast into the port as a drill-locating point to allow the customer to open up the intersecting Thermactor passages if he so desires.

Thanks to the revised port design, the initial evaluation by Kaufmann Products indicates that the new Dart heads offer a 20- to 25-percent-intake and an 18- to 20-percent-exhaust-flow improvement compared with the best production iron heads. They work at least as well as the aluminum offerings. Dyno tests recently conducted on an otherwise-stock 351W Ford Marine engine indicated significant over-2600-rpm torque and horsepower gains; the new Dart heads produced a virtually flat 335 lbs.-ft. of torque from 3100 through 3600 rpm—a 12- to 20-lbs.-ft. gain over the stock torque curve in the same rpm range. The horsepower peak increased from 229 at 4200 to 272 at 4900 rpm. Overall, the engine's average torque output increased from 292 lbs.-ft. to 316; its hp production increased from 198 to 220. All this, and you won't have to pay a king's ransom for these Windsors. As of October, 1990, the asking price for "Dart II Windsor By World Products" heads is only $639 per pair or $939 completely assembled. **HR**

SOURCE

World Products
Dept. HR
80 Trade Zone Ct.
Ronkonkoma, NY 11779
(516) 737-0372

RECOMMENDED COMPONENT PARTS NEEDED TO COMPLETE HEAD ASSEMBLY

DESCRIPTION	MFG	PART NO.	AR
Chevrolet intake valve, 2.02-inch*	Various	——	8
Chevrolet exhaust valve, 1.60-inch*	Various	——	8
Valve locks (11/32" x 7°)*	Various	——	32
Chevrolet valve-spring (1.25" o.d. x 1.700" installed height)*	Various	——	16
Chevrolet valve-spring retainer (1.25" o.d. spring and 11/32" valve stem)*	Various	——	16
Valve seals (for above valves and springs, type optional)*	Various	——	16
Chevy pushrod guideplates*	Diamond Racing	4230	8
Rocker arm studs*	Ford Motorsport	M-6527-A311	16
Rocker arm, nonrail, cast iron (w/fulcrum and nut)	Ford Motorsport	M-6564-A321	16
Pushrod (289/302W)	Ford Motorsport	M-6565-C332	16
Pushrod (351W)	Ford Motorsport	M-6565-B332	16
Head bolt kit, (stepped for blocks w/ 7/16-14 bolt holes) (*one kit needed per head*)	Ford Motorsport	M-6065-B289	2
Spark plug, tapered seat, 14mm x .472" reach	Various	——	8

*Supplied in World Products kit 7510

FLOW DATA

Data supplied using 28 inches of water pressure drop on a SuperFlow flow bench. Data represents the percentage of flow from a 300-cfm orifice (300 cfm = 100 percent), while numbers in parenthesis are actual cfm of airflow.

	WORLD PRODUCTS DATA		KAUFMANN PRODUCTS DATA	
VALVE LIFT	2.02-INCH INTAKE STOCK	1.60-INCH EXHAUST STOCK	2.02-INCH INTAKE BOWL BLEND	1.60-INCH EXHAUST BOWL BLEND
.200	42 (126)	30 (90)	41 (123)	36 (107)
.300	56 (168)	37 (111)	57 (170)	45 (134)
.400	68.5 (205.5)	41 (123)	71 (214)	49 (148)
.500	74 (222)	45 (135)	80 (241)	52 (155)
.600	77 (231)	47 (141)	86 (257)	53 (159)
.700	79 (237)	48 (144)	86 (259)	—(—)

No-Blow 350 hp

Build 350 hp Without A Power Adder

text by Chris Hemer
photography by Chris Hemer and Ford Motorsport

At a time when power adders rule supreme—when a new supercharger, turbocharger, or nitrous kit is developed almost every month in an on-going effort to satisfy the hunger of the 5.0 marketplace—there are still purists among us who will not accept a horsepower figure unless it came from a normally aspirated engine. They reason that using a power adder is cheating; anyone can make horsepower if they strap on enough artificial aspiration. These are the guys who stay in the grand stands to watch the Competition Eliminator cars run while everyone else goes to get hot dogs. They follow Pro Stock cars through the pits just so they can get a whiff of 116-octane racing fuel. And, they're the guys who will stop at nothing to build a streetable 5.0 that will make strong horsepower without the aid of the coveted centrifugal blower or the big blue bottle.

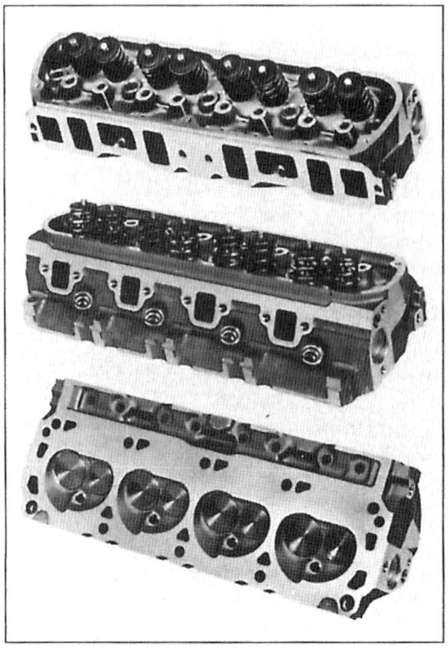

If you're one of those purists, then this article is for you. We recently combed the country and solicited expert tips from some of the most respected names in the industry in an effort to come up with truly streetable combinations that can produce 350-plus hp with a stock short block, normally aspirated.

Considering that your 5.0 made 225 hp right off the showroom floor, you may not consider another 125 hp to be that hard to produce—and it isn't. However, selecting the correct combination to yield this kind of power, yet still be completely reliable and streetable, can be a little more difficult, and requires some input from the people that know the 5.0 inside and out. To this end, we've talked to Kevin Peterson at JBA Racing in San Diego, CA; Lidio Iacobelli from Alternative Auto Performance in Mt. Clemens, MI; and Joe Amato from Downs Ford in Toms River, NJ. To satisfy the needs of the 4.6 contingent, we also ventured into the Great White North to talk to Sean Hyland of Sean Hyland Motorsports in Ontario, Canada, to get his tips on modifying the modular motor. Finally, we asked each participant to provide us with any other parts and/or pieces to complement such an engine, such as fuel system upgrades, gear ratio recommendations and the like.

As you may have guessed, no two people we interviewed had the same method of arriving at 350 hp, but you can be sure that every one of these combos will work.

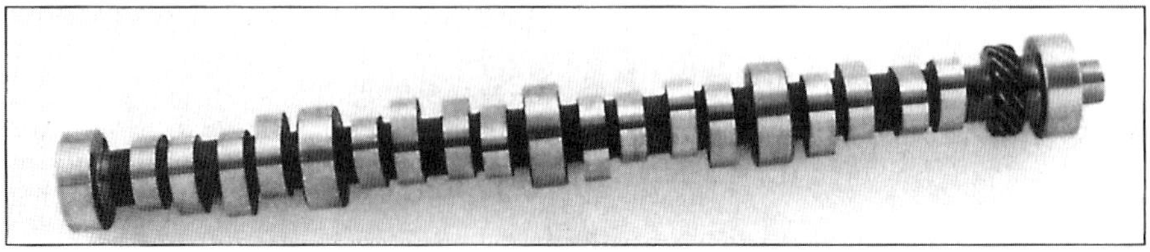

No-Blow 350 hp

KEVIN PETERSON—JBA RACING

JBA has been in the business of building 5.0 engines and complete cars since 1985, and is legendary in the Southern California area. In addition to their famous short-tube headers, JBA also features a well-stocked speed shop, and a service center staffed by seasoned professionals. JBA's service manager, Kevin Peterson, is known as "Mr. Horsepower," and has dozens of street-friendly engine packages that can suit the needs of any customer who walks through JBA's doors.

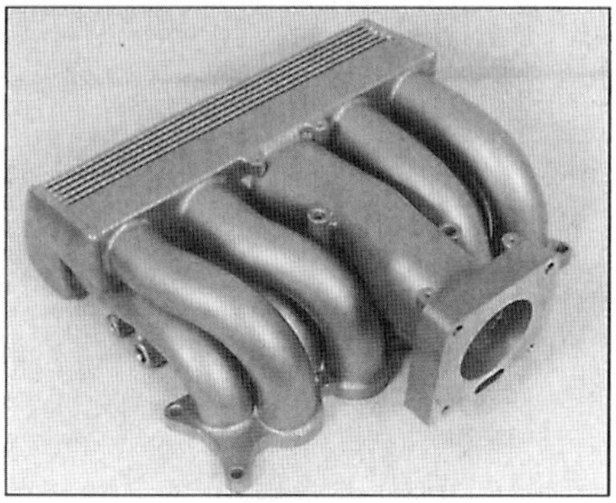

KEVIN'S COMBO:

INTAKE(S): Edelbrock Performer
Comments: "Several intakes flow well enough for this horsepower level. The Holley SysteMax II, Edelbrock Performer or Motorsport Cobra intakes will all get the job done. My personal preference is the Edelbrock Performer, due to the superior design of the lower manifold. The runners afford a much straighter shot at the intake valve than any of the others, which can promote better throttle response."
THROTTLE BODY: Accufab 70mm
Comments: "The Accufab unit is my choice, but the Motorsport 65mm is also a good piece."
INJECTORS: 24lb
Comments: "The minimum required is 24lb, but don't be afraid of 30-pounders. Unless a blower or a big nitrous system is in your future, 30-pounders are it."
MASS AIR: Pro-M 77
Comments: "I've tried them all, and there's still none better than the Pro-M 77. If your budget is tight, however, use a Pro-M 75 Bullet."
CYLINDER HEADS: Holley SysteMax
Comments: "Holley SysteMax heads are my first choice due to their bulletproof design and ease of installation. They will make 350 hp out of the box, but work even better with minor bowl work and a port match. Other heads will get you the power that you want, but none will live longer. Also, most people don't want to adjust their valves regularly. These heads use bolt-down rockers which require no maintenance."
CAMSHAFT: Lunati #51014
Comments: "The cam is the single most important component. I recommend the Lunati #51014. It's got 218-degree duration intake, 226-degree exhaust at .050, with .500 lift intake, .510 exhaust. This cam works great and will still be driveable. You can get more horsepower with more cam, but 99 percent of drivers can't deal with the driveability quirks. We also use a 6.400-long push rod with the Holley heads and the Motorsport 1.6-ratio bolt-down rocker arms."

HEADERS: JBA 1⅝-inch "Shorty"
Comments: 'nough said.
OTHER RECOMMENDATIONS: 180-degree Motorcraft thermostat; 190-lph Holley in-tank fuel pump; Motorsport underdrive pulleys; K&N air filter; Paxton adjustable fuel pressure regulator (baselined at 40 psi); TPS adjusted to one volt, initial timing, 12 to 14 degrees using 92-octane pump gas. "We also recommend an aftermarket shifter and a pair of weld-in subframe connectors," says Kevin, who also offers this tidbit of advice: "Use 3.73 gears, if you've got a pair—3.55s if your wife keeps them in her purse."

Source:
JBA Racing
Automotive Engineered Products, Inc.
Dept. 5.0
7149 Mission Gorge Rd., Suite D
San Diego, CA 92120
(619) 229-7797

LIDIO IACOBELLI—ALTERNATIVE AUTO PERFORMANCE

Alternative Auto Performance may not be a household name, but in eight years Lidio Iacobelli has developed a loyal following in the Motor City area. That's because he has his finger solidly on the pulse of the 5.0 market, and knows exactly what it takes to fulfill a customer's goal—no matter how big or small that goal might be. As a grass-roots racer himself, Lidio focuses on proven performance combinations that don't cost his customers an arm and a leg. A host of 9-, 10- and 11-second Mustangs provide mute testament to his abilities.

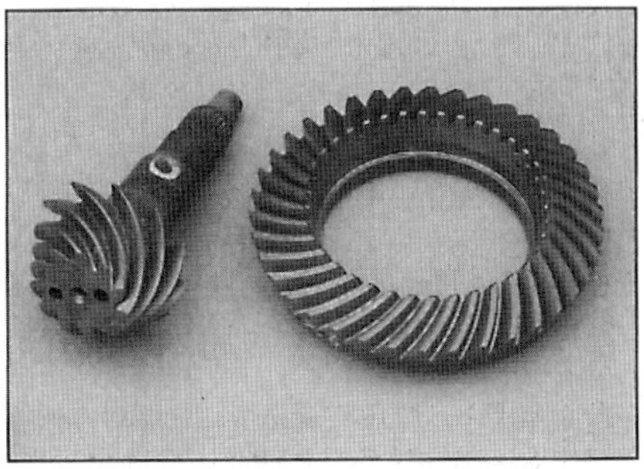

LIDIO'S COMBO:

INTAKE(S): GT40/GT40 lower, Cobra upper
Comments: "We use the GT40 stuff almost exclusively—either GT40 upper and lower, or Cobra upper and GT40 lower. The Cobra upper is far more popular right now, so we keep ported GT40 lowers in stock at all times for trade-in.

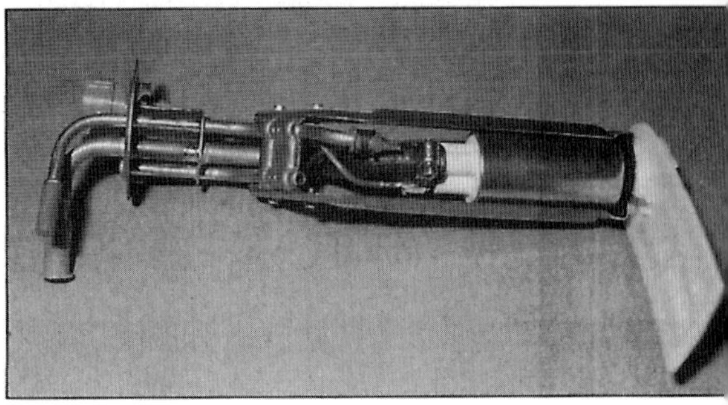

We've found porting them is good for about 10 to 15 hp; the upper doesn't require much work at all for an application such as this."

THROTTLE BODY: Motorsport 65mm
Comments: "I never go bigger than 65mm on a 5.0. There's no need for it—even with a blower. I use the Motorsport throttle body. I think it's the highest quality out there."
INJECTORS: 30- to 36lb
Comments: "Most engines in this power range will do really well with a 24lb injector. But nowadays, I usually talk the customer into 30- or even 36lb injectors because, at some time, they almost always go to a blower. 24lb is definitely the minimum for a 302 making 350 hp."
MASS AIR: 73mm C&L or Pro-M Bullet
Comments: none
Cylinder Heads: TFS Twisted Wedge
Comments: "We fully port the TFS heads and deck them upwards of .040 inch to increase the compression ratio to 9.5:1. Then we fit them with Motorsport 1.72:1 'blue' rocker arms, ⅜ stud mounted."
CAMSHAFT: Stock or "Baby" Crower, #15510
Comments: "I have always been a proponent of running the stock cam and a good set of heads. A lot of guys feel like the have to change the cam to make big power, but at the same time they don't want to lose too much driveability. The Baby Crower almost mirrors the stock cam in duration at .050, but the lift is .531 with a stock-ratio rocker, .570 with 1.72 rockers. It works just like a stock cam with regard to driveability and low-end torque, but it raises the power range to 6,000-plus rpm with good heads."
HEADERS: MAC 1⅝-inch long-tube headers
Comments: none
OTHER RECOMMENDATIONS: Cooler thermostat; reduction pulleys; colder plugs (2 heat ranges below stock); fresh timing chain/gearset; 155-lph in-tank pump; Kirban adjustable fuel pressure regulator; phenolic spacer (when using GT40 upper only). "When a guy is looking to squeeze every drop of power out of his combo, we also recommend an electric fan. These are worth 10 to 15 hp, although they can cause a car to run hot or overheat in summer traffic," says Lidio. "We would recommend a 3.73 or 4.10 gear with our cam, plus slicks, no sway bar, HP Motorsports Megabite lift bars, front runners, and better shocks/ struts. A combo like this should be able to run 12.20 easily with a good driver."

Source:
Alternative Auto Performance
145 Malow St., Unit D
Dept. 5.0
Mt. Clemens, MI 48043
(810) 463-0010

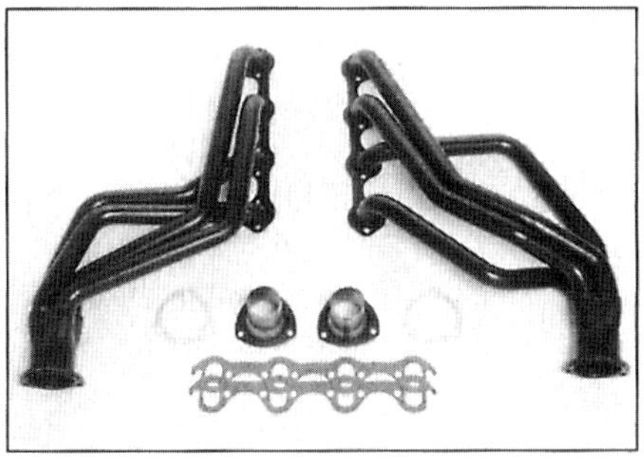

JOE AMATO—DOWNS FORD

There are a lot of Ford dealerships that sell Ford Motorsport products, but few are as dedicated to the 5.0 movement as Downs Ford in Toms River, NJ. Aside from being a Top-10 dealership for the last eight years, Downs Ford is the number-one Motorsport dealer on the East Coast and has featured a fully stocked Motorsport department since 1989. According to Downs Ford Motorsport department manager, Joe Amato, Downs doesn't do its performance work on the

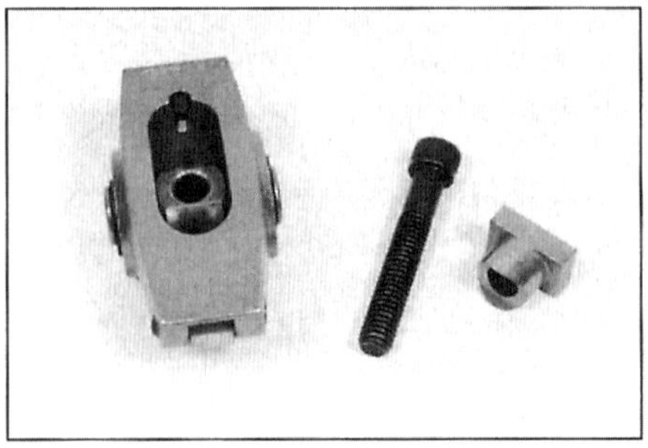

premises, but is hooked up with a full-service machine shop, chassis shop and installation center which can accomplish anything a customer desires.

JOE'S COMBO:

INTAKE(S): GT40 upper and lower
Comments: Port match lower manifold to heads
THROTTLE BODY: Motorsport 65mm and spacer
Comments: none
Injectors: 24lb
Comments: Use with Downs Ford adjustable fuel pressure regulator
MASS AIR: 75mm Pro-M Bullet is okay, 77mm preferred

Comments: none
Cylinder heads: GT40 X
Comments: "We recommend what we call a 'Bracket Port,' which includes full exhaust port work and bowl blend, but very little work on the intake side. Make sure you use ARP head bolts and Fel-Pro graphite head gaskets."
CAMSHAFT: Lunati #51014, .500/.510, 218/226-degree duration at .050 on 112-degree lobe centers.
Comments: none
HEADERS: MAC 1⅝-inch long-tube headers
Other Recommendations: 190-lph in-tank pump; reduction pulleys; 180-degree thermostat; Motorcraft spark plugs; 2.5-inch Flowmaster exhaust system; 3.73 gearing; Steeda lower arms or Southside Machine lift bars. "With the use of a good tire, the car should run down in the low 12s if the nut behind the wheel isn't loose!"
Source:
Downs Ford
360 Rt. 37
Dept. 5.0
Toms River, NJ 08753
(732) 349-2240, ext. 58

SEAN HYLAND—SEAN HYLAND MOTORSPORT

Sean Hyland Motorsport (SHM) is unique; unlike other shops that dabble in 4.6 and 5.4 Liter engines, SHM works only on the modular motor—both two and four-valve configurations. As a former IMSA Firehawk and rally racing competitor, Sean's background is mainly in four-valve OHC engines like the Eagle Talon and Ford Taurus. So when the Cobra was introduced in '96, he bought one and summarily tore it apart to initiate an in-house development program. "We never did much work on the 5.0 because it was primarily a do-it-yourself market," Sean explains. "But when the modular motor was introduced, I realized all that would change, and people would be looking for someone who could modify a modular engine." Initially Hyland focused on bolt-on items. But over the last 2½ years, he has developed other pieces (such as cams, rods, and pistons) that have allowed him to push power levels to over 700 hp in supercharged form. Sean says, "We've played with all kinds of combinations and we're still experimenting." He predicts the SOHC combo below will make right around 340 hp with the stock short block. With a set of pistons and better bottom-end pieces, though, he could twist the engine to higher rpm and therefore make more power—up to 400 hp at 7,000 rpm—normally aspirated.

SEAN'S COMBO:

INTAKE(S): Ford Motorsport, #M9424D46
Comments: None
THROTTLE BODY: Cobra-style twin 57mm throttle body
Comments: None
INJECTORS: 24lb
Comments: None
Mass Air: Pro-M or C&L calibrated for 24lb injectors
Comments: None
CYLINDER HEADS: Ford Motorsport, #M6049D46
Comments: "Fit the heads with larger 46.83mm intake valves and 35.88mm exhaust valves. Clean the bowls up and gasket-match to intake and headers. The stock head bolts are not reusable, so we recommend ARP head studs. They cost twice as much, but they can be re-used. Also, we recommend doing the head change with the engine out of the car, due to space considerations and the fact that having the engine out makes it easier to degree the cams."
CAMSHAFT: SHM Stage I—.495 lift, 230-degree duration at .050, 113-degree lobe centers.
Comments: None
OTHER RECOMMENDATIONS: Auto Specialties underdrive pulley kit; 3.73 gears; SHM 15/8- to 13/4-inch stepped long-tube headers and H-pipe; SHM custom-calibrated chip

Source:
Sean Hyland Motorsport
Dept. 5.0
RR 1 Langton, Ontario
Canada, NOE1GO
(519) 875-3135

Carbureted Performance

Just in case you think we're ignoring you carburetor-faithful, we're not. It's just that most of what applies to a 350hp injected engine also applies to a carbureted engine—with the exception of the induction system, of course. Any of the above principles will work in a carbureted engine. Weiand, Edelbrock, and others offer intake manifolds for the carbureted 5.0. As far as the carb is concerned, you're not going to get away with the stocker if you want 350 hp, but both the Carburetor Shop in Rancho Cucamonga, California [(909) 481-5816], and Jones Performance Fuel Systems in Huntington Beach, California [(714) 848-5500], can give the 4180 Holley carb new life.

400hp Ford 30

It Was a Lot Easier Than We Expected!

By Leonard Emanuelson

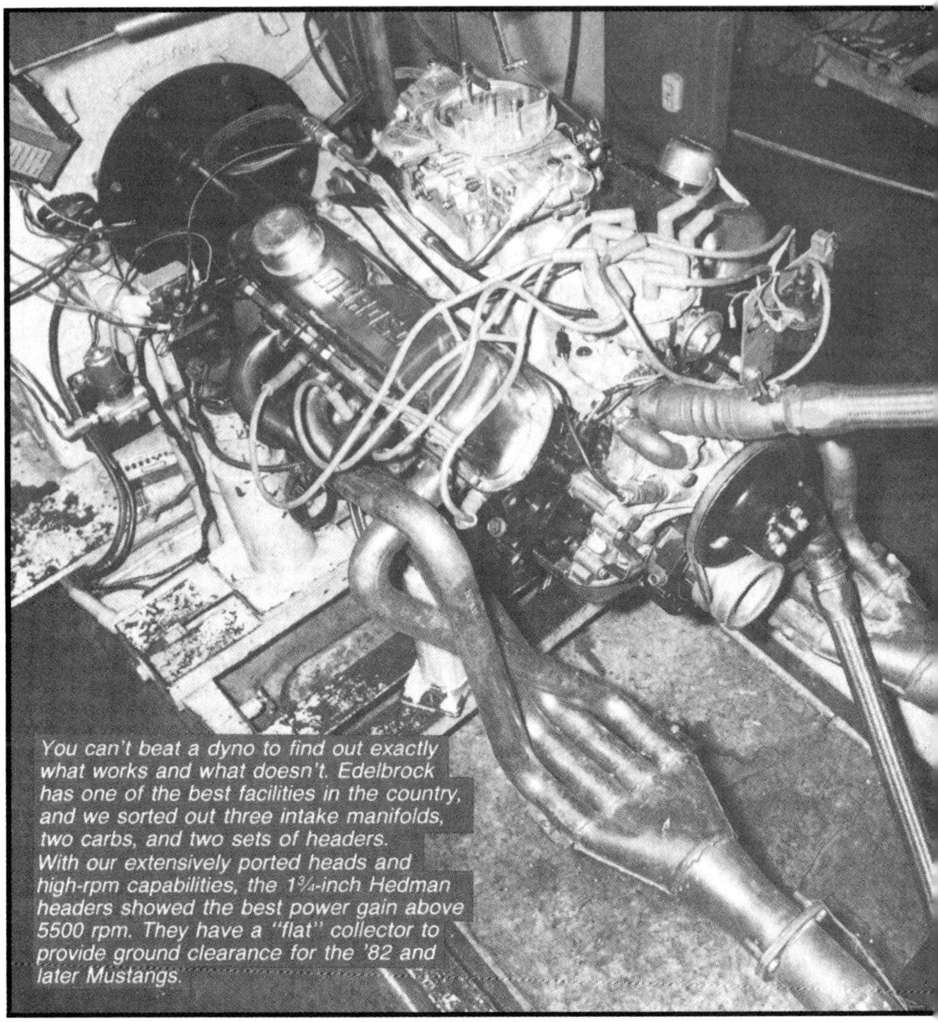
You can't beat a dyno to find out exactly what works and what doesn't. Edelbrock has one of the best facilities in the country, and we sorted out three intake manifolds, two carbs, and two sets of headers. With our extensively ported heads and high-rpm capabilities, the 1¾-inch Hedman headers showed the best power gain above 5500 rpm. They have a "flat" collector to provide ground clearance for the '82 and later Mustangs.

The challenge has been around the office for more than a year now. Let's pit the 305 Camaro against the 302 Ford and revive the classic Chevy versus Ford controversy. Although a hundred deadlines have been missed, the results are worth every bit of aggravation we've put into this research and development program.

And speaking of results, the 305 Chevy buildup was featured in the July '83 issue of HOT ROD. It produced 402 hp at 7200 rpm and a peak torque of 344 at 5500 rpm, all with bolt-on parts. Now that's pretty impressive, especially for a totally stock short-block smog motor. And because the Chevy was finished first, it really set the marks for us to shoot for with the 302 Ford.

Comparing the two engines is like comparing apples and oranges, even though Smokey Yunick has said many times, "The engine dosen't know what name is on the outside." It is a philosophical difference as much as anything, and once you analyze the differences you can get on with the building.

For emissions reasons, fuel economy, or whatever, Chevy started changing the bore/stroke combinations of their engines for smaller bore, longer stroke. Engines like the 302, 327, and 350 gave way to the 307 and the present 305. Many performance buffs thought that this was a mistake, but the new combinations made much better street motors with more low-end torque at a smaller displacement. Ford, on the other hand, has remained with the "oversquare" philosophy of large bore and short stroke, high-winding performance. So when comparing the Ford to the Chevy, you must understand that no matter what you do, the Ford will not match the Chevy's low end torque, so you have to build for high rpm power. It's strictly a matter of physics. A Chevy with a 3.5-inch stroke is going to make more torque than a Ford with a 3-inch stroke.

With that thought in the back of our minds, we took a look at the '82 H.O. 302 Windsor motor and formulated a plan of attack. Our approach to this whole program was to use bolt-on parts and keep the entire lower end stock so that the engine would not have to be removed from the car. Since this was the first Ford small-block that I had directed, I went back to what Smokey had said: it doesn't know what name is on the outside. Looking at the short-block you could say, "This is nothing more or less than a 302 Chevy—it has a 4-inch bore, a 3-inch stroke, 90-degree Vee, and a 90-degree crankshaft." One shortcoming over the Chevy is the fact that it only has four head bolts per cylinder to the Chevy's five, but being normally aspirated and with low-compression ratios this isn't a factor. It has a shorter length connecting rod than the 302 Chevy, which will help it make better low and midrange torque. In fact, the rod-to-stroke ratio is 1.69—a ratio similar to most big-block Chevys. The old 302 Chevy had a rod-to-stroke ratio of 1.9, and while this seems like a bunch of meaningless numbers it is one of the most important factors in designing an engine combination. Efficient cylinder filling is what making good power is all about, and a lower rod-to-stroke ratio motor has higher piston speeds (moves the piston away from top dead center quicker) and can make use of a much larger cylinder head port and a larger camshaft.

The next part of the engine to get our close scrutiny was the cylinder heads. Immediately, it became evident that if we were going to make some power, we needed a change of cylinder heads. While the H.O. GT cylinder head may be a good street head, we felt it was much too small in port volume for the performance we had in mind. Because of the rod-to-stroke ra-

W Buildup

tio, we determined that we were going to "over cam, over cylinder head, and over intake manifold" this engine. Referring back to the rat motor, have you ever looked at the size of the ports in a hi-po big-block Chevy? Also, past experience has shown us that the more cam you use, the harder a rat will run. Well, what we have here is a baby rat motor, and based on past experience, it was going to get plenty of port volume and camshaft—regardless of what anyone else said.

Although the Edelbrock Performer ***(top)*** *intake manifold made better low and mid-range power than both the Torker and the old dual-plane F4B, the Torker was the winner in the peak horsepower race.*

Small-block Fords use a single center cam bolt to attach the timing gear and fuel pump eccentric. That means the only thing driving the cam is the dowl pin, and when you start using offset bushing kits to degree the cam in, you're just asking for problems. Instead, we used a Pro Stock Engineering (16102 Orange Ave., Paramount, CA 90723, 213/630-4080) Pro-Dex timing gear set that uses a multi-keyed crank sprocket for advancing and retarding the cam.

Although Ford has built some incredible Cleveland cylinder heads, there is not much else available for the Windsor. We used a set of pre-'71 351 Windsor 4V heads that can be identified by casting numbers "C9OE" or "DOOE." They have smaller combustion chambers and bigger ports and valves than the '82 GT heads. Ford Power Parts (12200 E. Washington Blvd., No. C, Whittier, CA 90606) provided our castings which were then taken to Air Flow Research in Sun Valley, California, where Ken Sperling ported them within an inch of their life and installed Manley Pro Flow valves (see July '83 "Polished Performance").

Next we went for the *only* modification to the short-block, a camshaft. A few years back there were quite a few small-block Fords, namely 289s, that were getting Super Stock records around the country. A good many of them were using a Clay Smith flat-tappet grind camshaft. We decided to go with a proven cam, so George Striegle of Clay Smith (Buena Park, California)

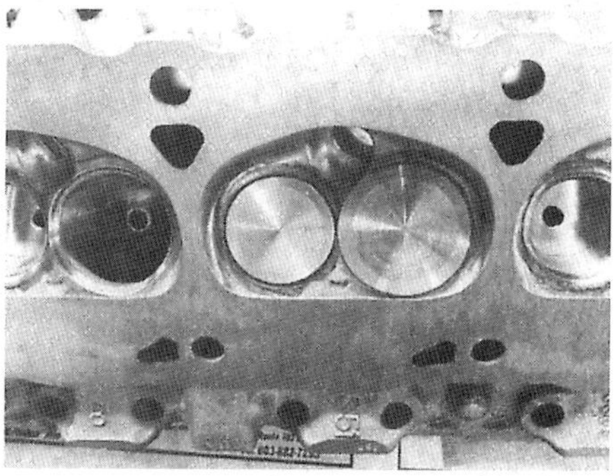

Our heads are 351W 4-barrel versions that have been ported by Air Flow Research. They use Manley Pro Flow valves and Clay Smith valvetrain parts with Harlan Sharp roller rocker arms. A future test includes a back-to-back test against the unported stock H.O. GT heads to see if the difference is worth the $2000 expenditure.

supplied us with a 302-4-B. It has 302 degrees duration at a checking clearance of .025, and .544 lift.

A camshaft this large doesn't just drop right in. It required deeper valve notches in the pistons, which we made in our garage with the engine in the car using Iskenderian Cam's piston notching tool. (That procedure was outlined in the May '83 issue in the "Cam-Rod-Ary" story, as was how we degreed-in the cam).

With the short-block and cylinder

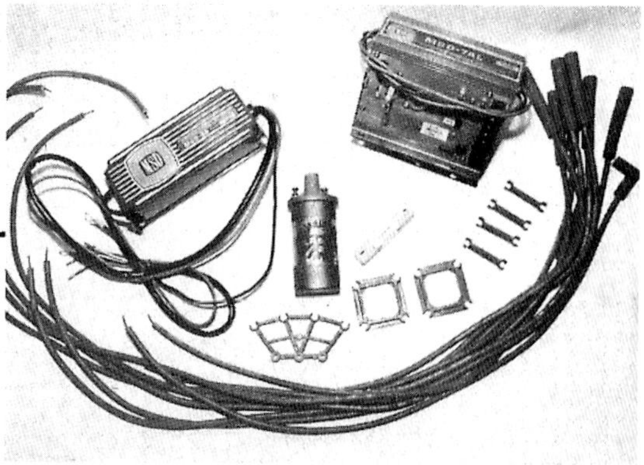

We discovered the real worth of the MSD-7A ignition system during the dyno test when the stock Ford ignition signed off at 5500 rpm. The MSD integrated with the Ford distributor eliminated the misfire and added considerable power at 6000 rpm and above.

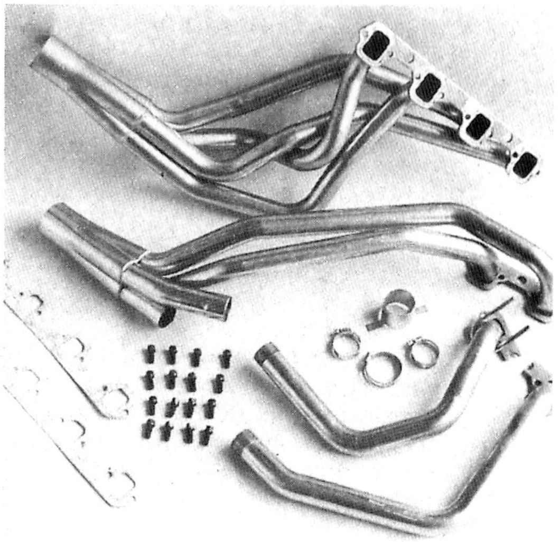

Hooker's four-tube, 1⅝-inch-diameter Mustang GT street headers do a good job. At 4500 rpm they are 25-hp stronger than the 1¾-inch headers. At 5500 rpm the larger headers take over, and at 6000 rpm they are 10-hp better than the 1⅝-inch headers.

heads handled, that left just three things to worry about—intake system, ignition, and headers. The 302W Ford is limited by what's available in the aftermarket. What we mean by this is that most headers and intake manifolds are on the small side because most Ford Windsor heads have small ports. So to start with we used Hooker 1⅝-inch (1½-inch i.d.) headers and an Edelbrock Performer intake manifold ported by Air Flow Research. The ignition system is a Ford distributor integrated with an MSD-7A box.

All of this theory and conglomeration of parts was thrown into our '82 H.O. GT Mustang and taken to the track while the majority of onlookers were tee-heeing and whispering "too much cam, those heads are too big, that thing won't fall out of a tree.... The Mustang had been fitted with a 3.8 rear gear and a C4 Rossi trans with a 4000

Photography: Marlan Davis

400 hp 302W

rpm stall converter. We surprised a couple of folks when the car proceeded to click off a 13-flat at 104 mph through the mufflers.

We used Fel-Pro gaskets exclusively in our Ford, and their performance was flawless. "Basic" overhaul set carries part No. FS-8548 PT-10, and includes all necessary gaskets and seals, including Permatorque head and intake gaskets. With our trick heads, we had to substitute '69-'74 351W intake gasket set MS-90116 for the intake gaskets enclosed with the kit. Although the kit's head gaskets proved satisfactory, we also received and used a then-experimental set of premium "Blue" gaskets (8883-PT); these are now standard high-performance items, but must be purchased separately from the basic overhaul kit.

To Blueprint or Not To Blueprint

Our intention was to do this engine modification as a bolt-on project, leaving the engine in the car, even for the valve notching procedure. That's exactly the way we did it, just like you would or could do it in your own driveway. Our engine had about 8500 hard miles on it before the modifications. After the modifications it got about another 1000 street miles and about 75 runs at the dragstrip put on it. All was well, until the oil pressure gauge showed a drop in pressure and there was a bad noise from the crankcase area. Yup, we had spun a main bearing!

The big question is, did it happen because of the modifications or would it have happened anyway. There is no way to tell, but after suffering a rod bearing failure in our 400 hp 302 Ford, we think a full disassembly and checking of clearances is needed if you're going to lean on these engines as hard as we have. Clay Smith Engineering (5870-F Dale Ave., Buena Park, CA 90621, 714/523-0530) blueprinted our stock shortblock the second time around, and Hank The Crank/HTC (7253 Lankershim Blvd., North Hollywood, CA 91605, 213/765-3444) blueprinted a crankshaft for us.

I guess it comes down to the old saying ". . . You can pay me now, or you can pay me later!"

Everyone was surprised and pleased, except me. I was looking for a 12.50 number, and after driving the car, I knew something was amiss. It had more bottom and midrange torque than it had any right to, and would fall on its face above 5800 rpm, just where the cam and cylinder heads should really start working. We put almost 50 passes on the car at the track in over a couple of weeks, changed a lot of things, but made no progress.

Jim McFarland offered the Edelbrock dyno, and dyno man Curt Hooker and Louie Hammel bolted our little Ford to the pump and proceeded to make a pull on it. I almost got sick when I saw the numbers—295 horsepower. "You've got to be kidding; my daughter can coax that number out of a Briggs & Stratton," I thought. Curt said, "But look at the brake specific fuel numbers; this thing's so fat (rich) it's a wonder that it'll run at all." He leaned it down about four jet sizes, went back, and made another pull. The torque beam reading was a high of 55 at 6500 rpm. After a few bleeps of the calculator it was established that we were making 360 hp. Much better. Just think, a 65 hp jet change. That's just how screwed up we really were at the track. What had happened was that we were running pump gas and, with 36 degrees of ignition lead, we had to keep richening up the carb to stop the detonation. We killed the detonation and the power along with it. It was a good lesson for everyone. *You can have all the right pieces, but if you can't tune the thing, they're worthless.*

With the knowledge that there was nothing major wrong with the engine it was time to get down to some old-fashioned dyno parts flogging. By the time it was all done, we compared two Holley carbs, three Edelbrock intake manifolds, and three sets of exhaust headers. Look to the horsepower curves for all of the particulars, but we will talk about a couple of combinations right now.

From past experience, we know that the bulk of an engine's power band falls within a 2000 rpm spread. That band can be moved up or down according to the cam, headers, intake, and carb. With the Mustang's 4000 rpm converter and 3.80 rear gear, we are looking for the power band to fall right between the 4000-6000 rpm range. Our final engine combination of 4776 Holley (600 cfm double-pumper), Torker manifold, 302 Clay Smith cam, and 1¾-inch Hedman headers gave us a torque curve that started out at 282 ft.-lbs. at 3500 rpm, moved up to a peak torque of 340 ft.-lbs. at 5500, 333 at 6000, and slightly tapered to 314 at 6500. Not bad for a little 3-inch-stroke street/strip motor.

Further, we confirmed some suspicions about the smaller tube 1⅝-inch Hooker street headers and the Edelbrock Performer intake manifold. The smaller 1⅝-inch street headers are about 30 ft.-lbs., or 20 hp, better than the 1¾-inch headers up to 5500 rpm, beyond that the larger ones are better, but only by about 10 hp and 8 ft.-lbs. of torque. The Performer manifold with a 1-inch open spacer is about 14 ft.-lbs. of torque and 10 hp better than the Torker at 3500 rpm, but the Torker gains superiority at 4500 and stretches the lead out to 15 hp and 13 ft.-lbs. at 6500.

This was one of the few cases where being conservative just didn't pay off. Fuel economy may be a different story, but when you're looking for power, you just have to go to the high-rpm combo.

Just to satisfy ourselves and the critics, we opted for one more camshaft change to prove that we hadn't overcammed this engine. Clay Smith ground us a shorter (C 292-8-B) cam that had 10 degrees less duration, less lift, and was ground on 108-degree lobe centers instead of 104. The engine lost 54 hp at 6500 rpm, quickly showing that this engine likes a lot of camshaft.

On the dyno we were 12 hp shy of the 400 we advertised, but since that test we have added a 50-cfm-larger carb that should put us over the crest. Our next stop is the dragstrip, and we'll find out for sure how well the little 302 Ford fares against the 305 Chevy and even those 426 Hemis of yesteryear. **HR**

Engine Dyno Test

RPM	HP	Torque
3500	188	282
4000	213	279
4500	257	299
5000	304	320
5500	357	340
6000	380	333
6500	388	314

Engine equipped with modified Edelbrock Torker manifold, 1-inch open carb spacer, 1¾-inch Hedman "flat collector" headers, Holley 4776 carb, and MSD-7A ignition system.

RPM	HP	Torque
3500	190	286
4000	230	302
4500	274	320
5000	312	327
5500	337	322
6000	346	303

Engine equipped with modified Edelbrock Performer manifold, 1-inch open carb spacer, 1⅝-inch Hooker street headers, and Holley 4776 carb.

Valvespring R&R

How to Swap Performance Valvesprings into Your Stock 5.0

By James Lawrence
Photography: James Lawrence

Changing your valvesprings can be fairly easy if you have the right tools for the job. Powertrain Dynamics often performs these valvespring swaps and had everything we needed to complete the job. Here Powertrain's Bill Bassett is compressing the stock weenie valvesprings and removing the stock valve locks with a telescoping magnet.

Despite the proliferation of affordable aftermarket aluminum heads for the 5-liter, there are plenty of enthusiasts still ripping around with stock heads. Problem is, when you start swapping on headers, H-pipes, 1.7 rockers, and GT-40 intakes, and even adding blowers or nitrous kits, the stock worn valvesprings become unable to control the valvetrain, a situation referred to as "valve float."

Stock 5.0 Mustangs don't come with the beefiest springs, retainers, or keepers to begin with. From the factory, 302 heads are equipped with a single coil spring with a damper and are adequate to control the valvetrain up to about 5,500 rpm. However, as the spring wears and as modifications are made to the engine that cause greater engine acceleration, the valvetrain's weight and inertia can overpower the springs at rpm above 5,000, which creates valve float.

Valve float can be very dangerous, beyond not being able to rev your engine as high as you might want. Often when a valve hangs open, it is smacked by a piston, bending the valve and sometimes even breaking it off and leaving it to flail around in the engine loose. This causes serious engine damage. At the very least, valve float will cause a massive loss of horsepower at whatever rpm it begins.

It happened that we were planning and preparing *5.0 Mustang*'s newest test mule, a 65,000-mile-old '89 coupe for the Bolt-On Dyno Thrash found on Page 107 of this issue. We figured that while we were under the valve covers swapping on 1.7 rockers, we should show our readers how to swap on some stiff performance valvesprings and high-quality valvetrain hardware. Yes, it can be done in an evening over beer and pizza. Competition Cams and Pacific Performance provided the parts, and Powertrain Dynamics did the wrenchin.

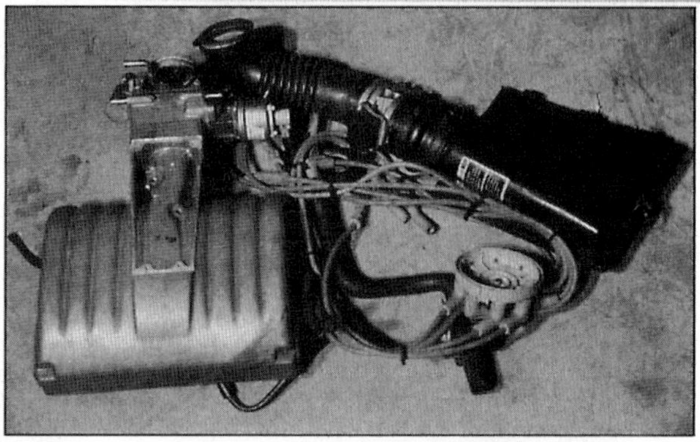

1 Powertrain started out our valvespring R&R by removing the upper intake manifold, distributor cap and wires, air boot, mass air, and air box assembly. You'll need to clear out all these parts to make enough room to remove the valve covers.

2 Our 5.0 Mule still has all its creature comforts (A/C) and smog compliance equipment (smog pump, EGR), so working around the engine can be a pain. Bassett's next step was to remove the valve covers.

3 With the covers removed, Powertrain unbolted the stock rocker arms. You should label each part (cylinders No. 1–8, intake or exhaust) and put it back in the same location. At this stage, you'll be removing the rockers and rocker bolts, the pedestals and stands, and the pushrods.

4 *Now you need an air compressor to help you pressure the cylinder with air so that the valve doesn't fall into the engine. Screw the spark plug/air hose adapter—available at most parts stores—into the spark plug hole of the cylinder that you will be working on (arrow). Powertrain's mechanic begins by compressing the coil using a Craftsman spring compressor.*

5 *We selected Competition Cams No. 972-16 single valvesprings with dampers that provide the desired spring rates for our application (125–135 lbs of seat pressure and 285–300 lbs of open pressure). We did not use the inner valvespring, which was included in the photo. Comp Cams also provided us with a set of their Steel 10-Degree heavy-duty retainers, 10-degree $\frac{11}{32}$-inch Super Locks, and a generous supply of valvespring shims. Many thanks to Pacific Performance for getting us, late at night, a pack of their P5052 valve stem oil seals.*

What Do You Need?

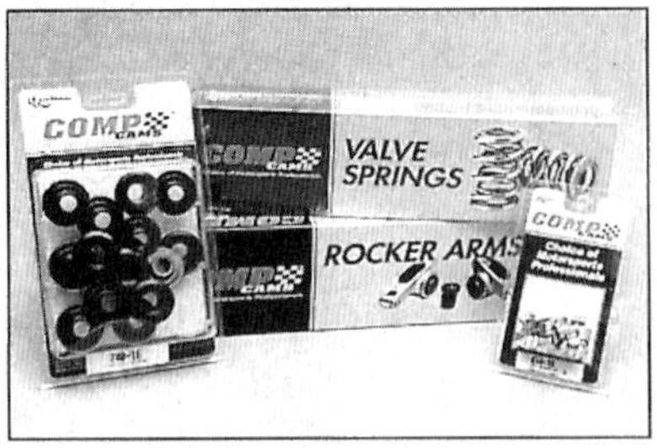

Competition Cams provided the valvesprings, retainers, and most of the other goodies for our test.

Parts

Competition Cams:
- 972-16 valvesprings, 1.460-dia.
- 740-16 1.437 steel retainers
- 614-16 +.050 $\frac{11}{32}$ super locks
- 613-16 $\frac{11}{32}$ super locks
- 4754 1.437 spring shim kit

Pacific Performance:
- P5052 Positive Stop oil seals

Tools
- Basic sockets and wrenches
- Valvespring compressor
- Dial calipers
- Air compressor
- Spark plug/air hose adapter
- Telescoping magnet

6 *Powertrain then measured for the installed height of the valvespring. Stock 5.0 and 302 heads typically come from the factory with different installed heights for the intake and exhaust, so you often have to use a combination of valve shims (available in .010, .015, .030, and .045 thicknesses) and +.050 valve locks to get the same installed height for both valves. Our target for the Competition Cams No. 972 springs, to get the proper open valvespring pressure, was an installed height of 1.750–1.780 inches. Powertrain used a .015-inch shim on the intakes and a +.050-thousandths retainer on the exhaust side—but your engine might be different, so be sure to check!*

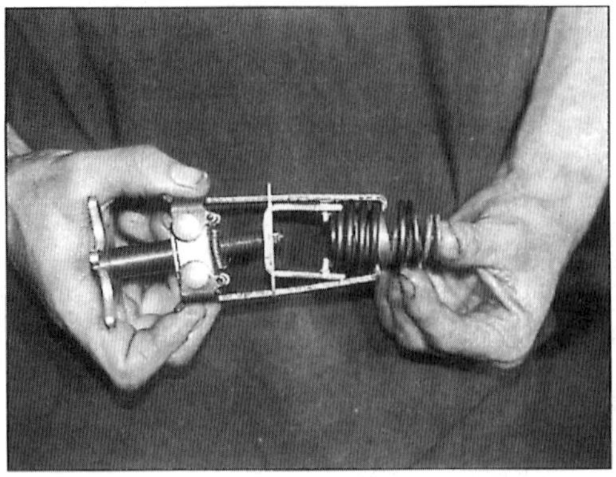

7 *After you've determined which valvesprings need shims and/or +.050 locks, go ahead and install the suckers. Our Craftsman compressor required us to mock the spring up by hand before installing it onto the head. Don't forget your shims!*

8 *The complete assembly (arrow). Now that's peace of mind when you're on the dyno for repeated 5,500-rpm full-throttle pulls!*

Sources

Competition Cams
Dept. 5.0
3406 Democrat Rd.
Memphis, TN 38118
901/795-2400
901/366-1807 fax
www.compcams.com

Pacific Performance Products
Dept. 5.0
1004 Orangefair Ln.
Anaheim CA 92801
714/773-4177

Powertrain Dynamics
Dept. 5.0
15628 Graham St.
Huntington Beach CA 92649
714/373-0068

POLISHED PERFORMANCE

A Ford Windsor Head That's the Equal of a Small-Block Chevy? Thanks, Air Flow

By Marlan Davis

Standard small-block Ford heads . . . miniscule ports, even smaller valves, an ugly Thermactor boss sticking right up through the exhaust port. They're born losers, right? You want a small Ford to run, go out and bolt on some Boss heads . . . and change the pistons and rods while you're at it (so it isn't really a bolt on, is it?).

Well, we're here to tell you there's another, easier way to go for all but the most radical competition cars: get a set of early 351W heads, stuff Chevy small-block valves in them, and *then* have Ken Sperling at Air Flow Research Heads (11754 Roscoe Blvd., #6, Sun Valley, CA 91352, 213/767-4270) give them the full port and polish. The result is a Ford Windsor head that will flow almost as well as Chevy straight-plug LT-1 heads. Now admittedly, heads of this quality aren't cheap, but have you checked out what Boss heads, pistons, and rods are going for these days?

Head Selection

The 1982 302 heads not only have small ports and valves, they also have large combustion chambers to help keep the compression down. There are some Windsor heads with larger ports, such as the old 289 high-performance heads, but they don't exactly grow on trees these days. That leaves 351W heads—they're fairly plentiful and have larger valves than any 289 or 302 Windsor head, including the 289 high-performance. But even 351W heads are not all created equal: the later the head, the more its layout is compromised for emissions purposes.

Ideally, you should be looking for heads with casting numbers "C9OE" or "DOOE," which are fairly plentiful in the salvage yards. Heads from '71 through '76 have larger combustion chambers, while the '76-and-up castings have smaller 302-size valves in addition to the larger chambers. Any head purchased new from Ford—regardless of what the catalog claims the current part number replaces—are the inferior late-model "smog" castings. Ford Power Parts (12200 E. Washington Blvd., No. C, Whittier, CA 90606, 213/921-5300) has a supply of the right heads on hand, and helped to put us one step a*head*.

Plan of Attack

While 351W heads may have larger valves than other Windsor family members, they're not in the same league as their small-block Chevy brethren, where 1.94/1.50 valves are as common as fleas on a dog and 2.02/1.60 high-performance combos are to be found by the thousands. Since our heads were going on an '82 Mustang shootout car (we'll be running off against a late Z28 with results in an upcoming issue), we knew they needed help in the valvetrain department. Manley Performance Products (13 Race St., Bloomfield, NJ 07003, 201/743-6577) was contacted and they shipped us a set of their new "Pro Flow" Chevrolet valves that feature standard Chevy 11/32-inch-diameter stems for durability, but neck down just above the head to 5/16 inch for improved flow.

With pieces in hand, Ken got down to business. After several weeks of work, he rendered his verdict, "The port shape is not really bad, it's just too small. The heads are designed for an 'econo-grocery-getter,' and represent 1960s technology. Since most guys run Cleveland heads, not much has been done. The Ford heads are so small, this is one case where bigger is definitely better. I just tried to apply some of the latest porting concepts. The biggest gains with any head are to be found by centering the valve in the pocket. The ports should be aligned to the center of the valve guide; the pocket transition should be circular. Try to straighten the walls, and all turns and bends should be a constant radius. As delivered, this, of course, is not the case with the Ford heads, so we try to correct it, but the water jacket interferes. Pushrod location, however, is not that big a problem." So, with heads, as in life, everything's a compromise.

Exhaust Port

As delivered, there is a pronounced "ski jump" in the exhaust port floor. This should be smoothed out as much as practicable. If so equipped, the Ford Thermactor (smog pump) boss must also be removed. Widen the port exit to obtain a slight mismatch with the header flange, using the header gaskets as a template. Ideally, to combat reversion the exhaust port opening should be slightly smaller than the header flange. On our heads, the exhaust port exit now measures 1.330 inches high by 1.180 inches wide.

Intake Port

At the port entrance, a slight anti-reversion mismatch is again required, but here the port needs to be slightly larger than the intake manifold. Ken achieved this by using the smaller 302 intake gaskets as a template for the intake manifold, while the heads were matched to the larger 351W gasket. Overall port dimensions are now about 1.230 inches high by .850 inch wide at the narrowest portion of the port where it squeezes by the head bolts.

Because of the oversize valves, and

This cutaway view illustrates Air Flow Research's porting philosophy as well as the ultimate limitations of the 351W heads themselves. Ideally, Ken tries to center the valve guide in the bowl, but the water jacket interferes. Since we were running oversize valves, the bowl area also required considerable enlargement. Once the bowl configuration is established, the port walls are widened and straightened to match the bowl's maximum diameter (again, subject to the casting's limitations).

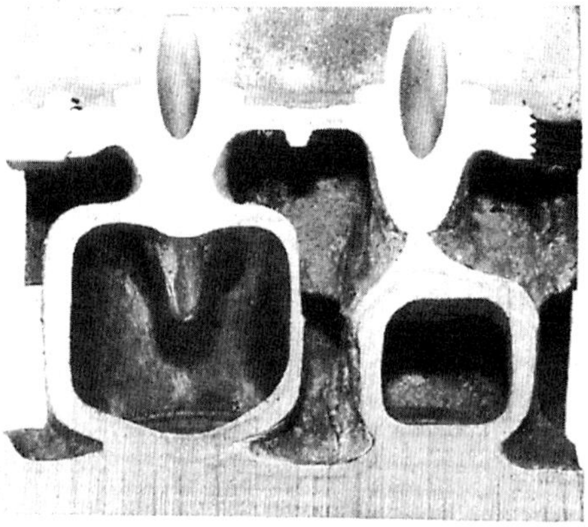

Another view of the valve guide area shows how far off the port centerline it is located.

The 1.94/1.60 Chevy-sized valves really fill the chamber—you literally can't get any bigger!

knowing this engine was being built primarily for off-highway use, Air Flow's Chris Brintnell really went to town on the intake port and, as can be seen from the photos, very little wall thickness remains in some spots. Going that far is obviously marginal if a 100,000-mile street engine is the goal, but it does maximize flow! Seriously, if extended operation is the goal, a minimum .080-inch wall thickness is recommended.

Combustion Chamber

The non-quench area of the combustion chamber was unshrouded out to match the engine's 4-inch bore size, using the supplied Fel-Pro head gasket as a template. While such a procedure is beneficial on any engine, it is especially critical in applications utilizing oversized valves so the engine may realize its full flow potential.

Since our stock 302 heads gave only 8.3:1 compression with the OEM flat-top eyebrow pistons (.0005-inch deck and

Not all 351W heads are created equal. The ones to get are cast "C9OE" or D0OE."

.047-inch head gasket), and we did not want to "re-piston" the motor, it was important to bring the combustion chamber volume down to minimum specs. Our heads were therefore milled .050 inch, reducing the volume from 65.8cc to 57cc. Present compression on the motor now specs out around 10.05:1.

Edelbrock's Performer intake is probably best for street use on the 289/302 Ford, but for our drag tests its performance was lacking beyond 6000 rpm—despite the fact that its ports were enlarged to mate with the larger 351W ports and the plenum was dropped ½ inch for improved top end.

If street/strip performance is your game, Edelbrock's Torker is the name. This single-plane intake offers more top end than the Performer, especially after Ken got through with it. Single-plane configurations like this one can easily be "flow balanced," or modified, so each runner will flow or displace as equal an amount as possible of air and fuel. Since all runners are not exactly the same length, a taller and/or wider cross section on the shorter runners is required to make up the difference. Naturally, all this is blended in with the port match and carb flange radius, as well as the usual general cleanup and deburr. We also have a 1-inch spacer for increased top end effectiveness, and Ken made sure it mates exactly with the as-delivered plenum.

Valve Job

With oversize valves, naturally the seats had to be enlarged. Ken prefers to rough the new seats in using a Winona seat machine since it's "clean, quick, and accurate. Stones won't stay sharp, and they're not as precise; the Winona gives a better center." The new seats are rough cut .030 inch undersize, which leaves enough material "to play with" for the final hand-ground valve job.

On the intakes, the seat face is cut at 45 degrees with a 35-degree angle on the combustion chamber-side and a 60-degree bottom cut. The transitional area from port to seat is critical. Ken says he uses a nominal 70-degree transitional angle, although in reality there is a continuous hand-ground arc of turn on the bowl radius.

Exhaust seats are similar, but the underside (throat) angle is altered to 55 degrees and radius-blended in the transition off the seat. A 22-degree backcut on the intake valve seems to help intake flow volume, but does not seem to aid the exhaust side.

Both intake and exhaust seats are .065 inch wide; while to some this may seem too narrow for continuous use, Ken replies, "This is what I do for Junior Johnson. I also do this seat in Prudhomme's Funny Car heads. In use, the seat automatically gets wider, so we're just compensating for the inevitable pounding the seat will undergo.

Intake Manifold

The intake manifold should really be considered an extension of the head's inlet port runner, since the best head

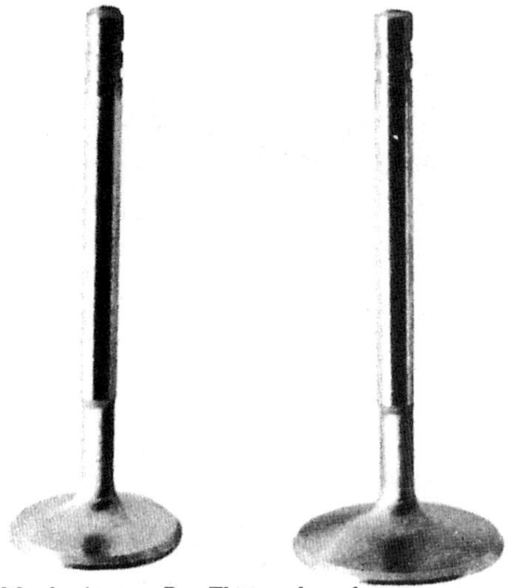

Manley's new Pro-Flow valves feature durable standard stem diameters, but neck down just below the head for enhanced flow. We used their part No. 11830 2.02-inch Chevy intake cut down to 1.94 inches, and a part No. 11827 1.60-inch exhaust. Both were also specially machined for use with Comp Cams' 10-degree keepers.

Air Flow attempts to apply modern precision techniques to the "art" of porting. For example, on common small-block Chevy heads, a computerized-tape machine does the rough-porting. If new oversize seats are needed (as they were on our Ford heads), they are roughed in with a precision Winona seat-cutting machine.

POLISHED PERFORMANCE

porting job won't work at its full efficiency if you can't get the air/fuel mixture there in the first place. And when it comes to small-block Ford intakes, unfortunately the selection is pretty limited. Since the engine is not normally used in high-rpm applications, there is no real strip—or even street/strip—intake available for it.

To date, we have tried two different Edelbrock intakes: the Performer 289 (Cat. No. 2121) dual-plane-style manifold and the single-plane Torker 289 (Cat. No. 2755). Both require considerable port matching since they are designed to work with stock 289/302 small ports. Ken modified the Performer for greater top end performance by dropping the plenum divider ½ inch, radiusing the carb flange, and in general enlarging the runners as much as possible, given the inherent interior runner accessibility problems of the dual-plane configuration. Our test indicated the manifold pulled well in the low-range and mid-range, but pooped out above 6000 rpm.

At the time this was written, we had not as yet track-tested the Torker (that'll also be in a future issue). Seemingly, it should be a better top end piece, but on the other hand, its runners are not so gigantic that all bottom end response will be lost. As seen from Ken's point of view, the Torker is easier to modify since, as a single-plane design with virtually straight runners, it is fully accessible top to bottom.

Valve Gear

Here, again, the Ford head's limitations must be taken into account. There is no room to machine the spring pockets for larger-diameter springs, while the spring pocket itself is rather shallow, with little room to add shims. Ultimately, if using Ford valves and serious racing cams, the valve stems themselves would need lengthening over stock, so the spring height in turn could be increased.

In our case, the Chevy-type valves are slightly *shorter* than the stock Ford pieces. To maintain the correct geometry, in theory .11-inch-shorter 289 pushrods are required. Therefore, Manley supplied us with their 6.776-inch-long hardened pushrods designed for use with guideplates, which in turn are required because the stock rail rocker arms and non-adjustable rocker studs were axled for Manley screw-in studs and Harlon Sharp roller rockers. Since the guideplates were .125 inch thick,

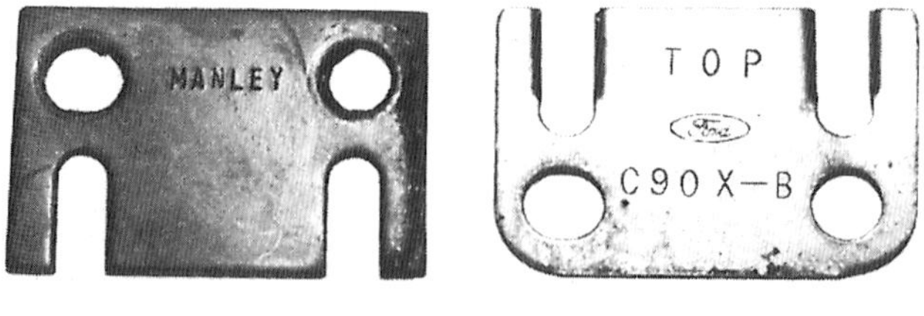

*You can't depend on "rail" rockers or slots cast in the head to control pushrods in a high-performance valvetrain. We have used both thick Manley (**left**, on engine) and thinner OEM-style plates supplied by Maier Racing. Manley supplied the screw-in rocker studs; Comp Cams, the Harlon Sharp roller rocker arms.*

Rocker stud bosses were milled .125 inch to compensate for guideplate thickness. Experience has shown that .250-.300 inch is a better figure for maintaining the correct geometry without going to odd-size pushrods. The old press-in studs were removed with a slide-hammer.

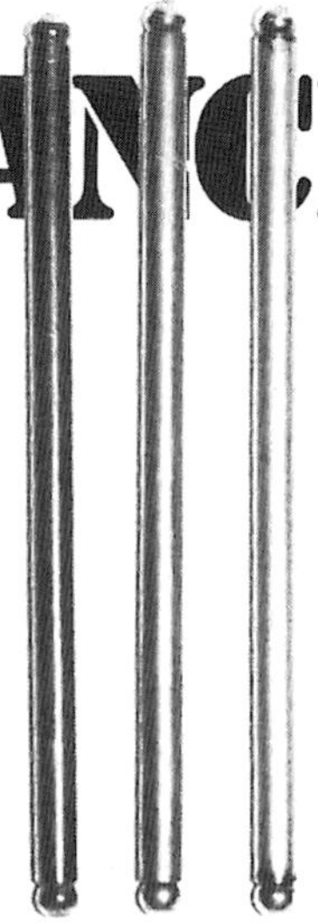

*Chevy valves require a non-stock-length pushrod since they are slightly shorter than Ford valves. In most cases a 289 pushrod **(left)** will do the job. However, we discovered that with high-lift cams and roller rockers, the rockers' Polylocks will bottom against the top of the stud. This was corrected by using .070-inch-longer (than stock 289) pushrods **(right)**. Stock non-hardened 302 pushrods **(center)** can not withstand rubbing against the guideplates.*

To maintain positive oil control, bronze wall guides were installed using a K-Line guideline tool, then bored out .060 with a self-tapping reamer. The top of the guides were also lowered .100 inch so they'd accept positive-type seals.

Stiffer Clay Smith valvesprings offer 145 pounds on the seat (at 1.800 inches installed height), with 350 pounds open. Also added was a spring cup so the springs would be positively indexed both top and bottom.

before tapping the stud hole Air Flow removed the equivalent amount of material from the stud boss' top.

As progressively more radical cam profiles were tested, we found the Polylocks were bottoming out on the stud top after only engaging three threads. This problem is correctable by several different methods or a combination thereof (valve stem lash caps, longer pushrods, longer Polylocks, longer rocker studs, and/or thinner guideplates). In our case, Maier Racing (235 Laurel Ave., Hayward, CA 94541, 415/581-7600) supplied us with his .070-inch-longer pushrods and .030-inch-thinner OEM-style guideplates. This combo solved our geometry problem, although milling the stud boss a total of .250-.300 inch would have done the job as well.

Competition Cams (2806 Hanger Rd., Memphis, TN 38118, 901/795-2400) supplied the initial single springs with dampner and steel retainers. Installed at 1.800 inches, they featured 120 pounds on the seat, with 330 pounds over the nose. Performance was satisfactory with hydraulic cams, but when a .545-inch Clay Smith Engineering (5870 Dale Ave., Buena Park, CA 90621, 714/523-0530) was tested, valve float occurred at 6300 rpm. We then had the "step" removed from the spring pocket so a Clay Smith spring cup and stiffer dual-spring setup could be installed.

However, despite what are really minor teething problems encountered with any hot combination, there is no doubt that these heads really work. The accompanying flow chart shows that the so-called "inferior" Ford castings flow nearly as well as Ken's conventional straight-plug Chevy LT-1 castings. Later, on an independent flow bench, these same Air Flow-prepped heads outflowed our erstwhile Z28 shootout competitor's heads nearly *two to one*. 'Nuff said! **HR**

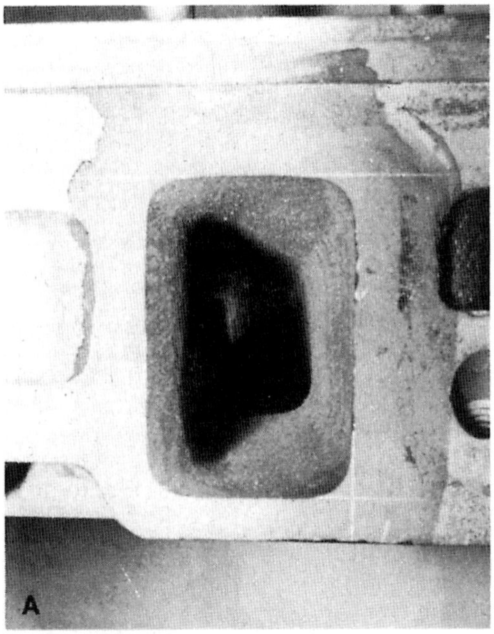

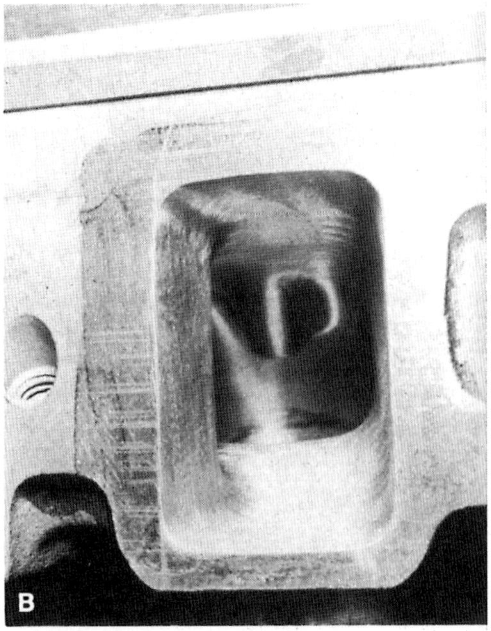

Before and after: turning a mongrel into a thoroughbred. Note how the stock intake port ***(A)*** *was straightened to a free-flowing "line-of-sight" configuration* ***(B)****. The exhausts* ***(C)*** *were likewise heavily massaged* ***(D)****—good-bye Thermactor boss!*

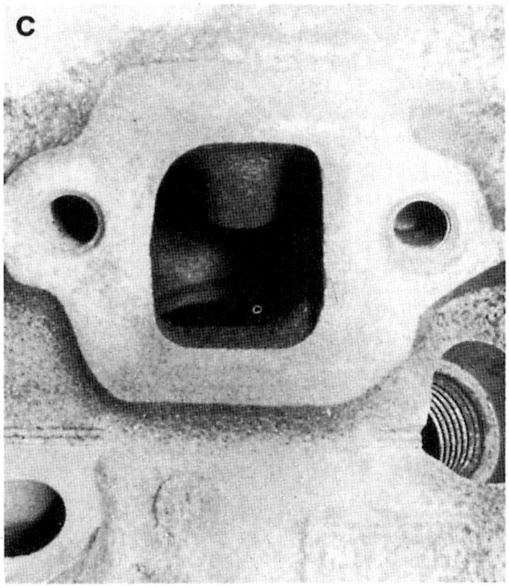

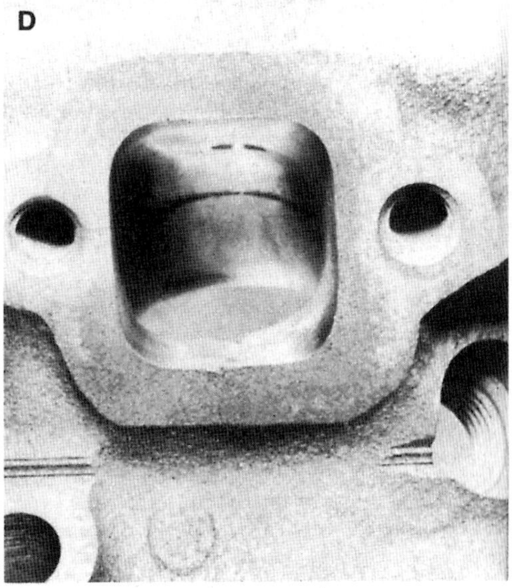

351 Windsor Ford Heads vs. Chevrolet Small-Block Heads

Our heads may be among the best 351W castings ever ported—just look at these flow chart numbers! The 351W ended up not comparing unfavorably with LT-1 castings, despite the latter's slightly larger 2.02-inch intake valve. Of course, there's no comparison with the latest "bow tie" castings, but those were designed with more than a decade additional experience under the engineers' belts.

All heads ported by Air Flow Research at 28 inches water drop, 300 cfm orifice. Flow numbers represent the percentage flow obtained to the total available capacity of the orifice (300 cfm).

Valve Lift	Ford		Chevy Straight-Plug LT-1		Chevy "Bow Tie"	
	Intake (1.94)	Exhaust (1.60)	Intake (2.02)	Exhaust (1.60)	Intake (2.05)	Exhaust (1.60)
.200	40	33	45	37	46	41
.300	60.5	45	65	48	66	53
.400	73	48	76	53	81	63
.500	75	50	78	56	90	66.5
.600	76	51	79	57	95	69
.700	—	—	—	—	98	—

PACE-MAKER

BY SWAPPING FOMOCO PAR
YOU CAN TRANSFO
YOUR '79 MUSTANG
INTO A JACK ROUSH-ST
RACE HO

By Gray Basker

Assembling an Indy pace car package 10 years ago was duck soup. It was as easy as calling up one of the Big Three and having them send you one of their big-block options. Oh, but how times have changed. We're thinking small, if not miniscule, these days, and finding something that hauls is like getting next to a girl whose father owns a liquor store and whose brother runs a chrome shop. That's the situation Ford discovered when they offered to supply the Indy 500 folks with their 1979 pace setters. What was needed was some serious horseflash coming out of turn four. Ford had originally intended to run their turbocharged 4-bangers, but production scheduling difficulties made the total number of units required impossible to obtain. Consequently, Ford decided to substitute inches for boost, and they eventually fell by Jack Roush's "power house" for some serious horsing around.

Normally, modifying a stocker for Indy has about as much street appeal as restoring a 16-valve Fronty. But this bit is different. The engines going into the three pace cars are mildly massaged 302s that are reasonable, Rock-of-Gibralter dependable and, most assuredly, af-FORD-able. They're built entirely out of over-the-counter FoMoCo and Jack Roush performance parts; they are docile enough to run around the pits; and they have the oats to possess 130-mph power. What's more, we conned Jack Roush into showing us how he transformed these '79 smoggers into 300-horse throwbacks.

A CHIP OFF THE NEW BLOCK

The easiest thing here would have been for Roush to slide in a 351-inch Windsor and pass it off as a 5-liter V8. But Ford would have none of that. So Jack bolted down the torque plates on a stock '79 302-inch block and passed the CK-10 down the bores with a No. 820-grit hone to straighten out the surfaces as well as to ensure a good seal. Additional machining operations included surfacing the decks, align-honing the mains, and grinding off the cast flashing to prevent any potential cracks or fractures. One thing that Roush didn't do was to remove the cam bearings. Evidently Ford inserts semi-finished cam bearings in the blocks and rebores them in place—and if you want to stay out of trouble, don't mess around with the cam bearings.

STROKING ALONG

Like in our "Small Wonder" yarn (October 1978 HOT ROD), Jack selected a forged steel Boss 302 crank (No. D0ZZ-6303-A) and armed it with eight 5.155-inch Boss 302 rods (No. C9ZZ-6200-B), a set of ⅜-inch rod bolts (C9ZZ-6514-A), nuts (C9AZ-6512-B), fully grooved TRW Clevite 77 main bearings (No. MS2975P), TRW Clevite rod bearings (No. CB743P), eight flat-top TRW replacement pistons (No. L2305F standard) and a ring combination made up of a Speed-Pro replacement ring set that includes a moly top ring, a cast-iron second ring and a low-tension oil ring (No. R9343 standard). Basic clearance figures are as follows: .002- to .002½-inch on the mains and rod bearings, .0006-inch pin clearance, and .004-inch piston-to-wall clearance. The longer Boss 302 rods made it necessary to take .040-inch off the top of each piston, and the skirts were notched so they would clear the crank's counterweights.

CAM SHAFTING AND VALVE TRAINING

In the vernacular of the hot rodder, the camshaft or "bumpstick" is an old rough-'n'-rasty muscle part from the Boss 302 days. The stick, known as part No. D0ZZ-6250-A is a solid-lifter cam that features a .477-inch lift on both intakes and exhausts, 290 degrees of duration, and is ground on 116-degree lobe centers. Instead of relying on the '79 Ford stamped rocker arms, "Cactus Jack" switched to the old-style rockers (No. C2OZ-6564-A), which use fully adjustable nuts (No. C2DZ-6A529-A) on the top of each rocker arm. In conjunction with the rocker arms are 5/16-inch chrome moly pushrods (JR6S-6565-L) with oil restriction on the top ends, a set of Jack's guide plates (No. JR5C-6A564-A), screw-in studs (No. C3OZ-6A527-B) and valve springs (No. C9OZ-6513-A). The springs, from a stock 351 Windsor, provide 90 pounds of seat pressure closed, 250 pounds open. Valving remains stock '71-to-'72 351 Windsor and features 1.84-inch intakes (No. D2AZ-6507-A) and 1.54-inch exhausts (D2AZ-6505-A), although Jack

PHOTOGRAPHY: GRAY BASKERVILLE AND FORD MOTOR COMPANY

What makes hot rodding so much fun is that you're always finding something unusual. Take "Cactus Jack's" engine test stand, for instance. It just goes to show you what a couple of skids, a pair of DieHards and some old hose can do. And you should hear that throttle response; it's a clean-running little sweetheart. Obviously, the stock 302-inch Ford cylinder block has been completely machined, and that includes align-honing the mains, resurfacing the decks, cleaning up the cylinder bores .005-inch, rounding off all the sharp edges, chamfering the lifter bores, and retapping all the bolt holes.

has been toying with TRW Chevy replacement valves (1.940-inch intakes, 1.600-inch exhausts) that are larger than the 351s.

FLOW JOBS

While the Original Equipment Manufacturer cylinder heads are quite adequate for today's performance standards, they won't flow enough air and fuel to make the kind of power deemed necessary for Indy pace car requirements. So Jack changed to a pair of '71-to-'72 351 Windsor heads (No. D1OZ-6049-B) that will bolt directly to the '79 302-inch Ford block. Additional machine work to the heads included milling .080-inch off the bottom surface to decrease the size of the combustion chambers to 51cc and to bring up the static compression to 10.9:1.

INDUCTION

Although Ford offers a fine 2-barrel aluminum intake manifold on their new 1979 302 V8, and even a 4-barrel cast-iron model for the vintage 289 high-performance engine (No. C4OZ-9424-H), Roush selected the Shelby & Marine 4-barrel aluminum intake (No. C9OZ-9424-D), which was good for an additional 25 horsepower over the stock Ford 2-barrel package. To better ensure a proper fuel mix, Jack installed a Holley No.

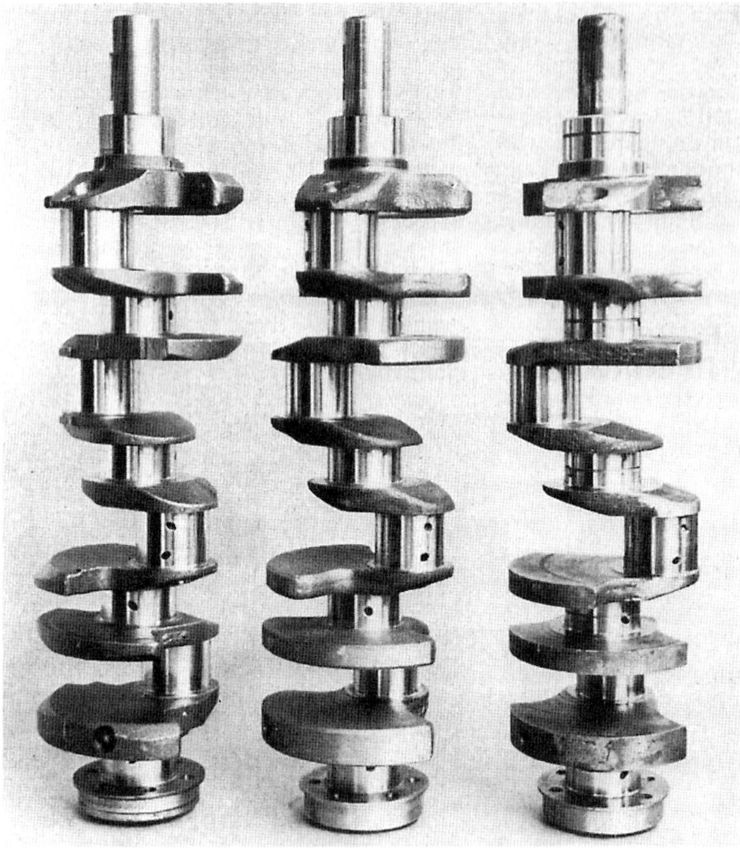

The new 302 will take any one of these cranks. Left is a stocker, center is the forged Boss 302 (with a 3-inch stroke), and the one on the right is a forged 292 truck model with a 3.3-inch stroke.

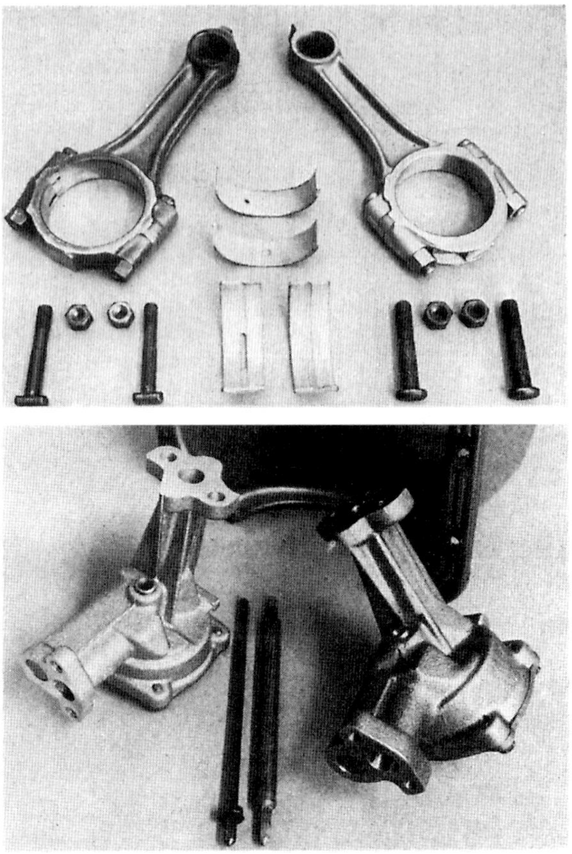

Stock rod and 5/16-inch bolts (left) were swapped for the ⅜-inch-bolt-equipped Boss 302 rods. Big Roush pump was used instead of stocker on left.

Piston (left) is a stocker, and the TRW 289 (middle) has been brought up to 302 specs. Indy version has narrowed skirts (arrow) to clear counterweights and notched valve pockets for 351 Windsor heads.

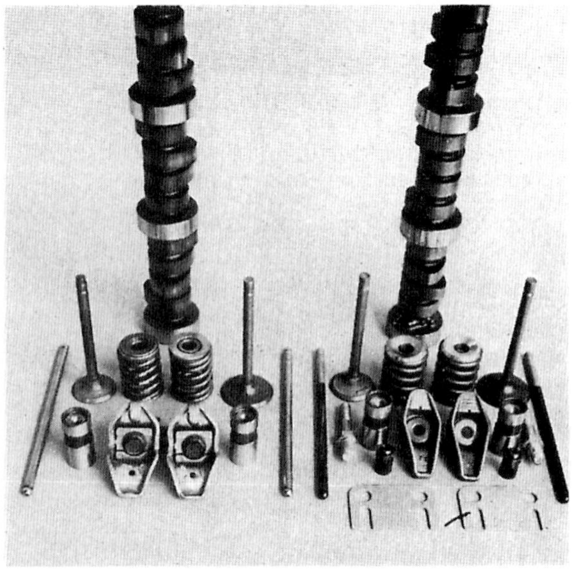

Muscle-type hydraulic (left) can use stock Ford valvetrain. Jack used the Boss 302 mechanical with old-style adjustable rockers, chrome moly pushrods and guide plates.

R6989 600-cfm carburetor that features an electric choke and vacuum-operated secondaries. Jack further modified the fuel delivery curve by putting in No. 35 power valves and No. 64 primary jets, drilling out the high-speed air bleeds to .035-inch, installing .029-inch primary idle bleeds, opening up the secondary metering plates to .076-inch, and richening the secondary idle circuit by going from .028-inch to .031-inch idle feed restrictors, then topping it all off with a Ford police-type air cleaner (No. D4AZ-9600-B).

IGNITION TO EXHAUSTION

No exotica here, because all Jack did to the stock distributor was to recurve the advance to Boss 302 specs and install a set of Autolite AF-22 spark plugs gapped at .035-inch. And the exhaust system is factory original except the catalytic converter has been removed for this off-highway, emission-exempt use.

Jack figures that a true blue Ford-natic could duplicate his Indy engine effort for something less than three grand. He is of the opinion that you could come down on the compression ratio a point and a half and end up with a 5-liter Ford that would put out an honest 270 horsepower at 6000 revs. Now when you stop to think about it, adding a safe and sane 140 ponies to your '79 'stang isn't all that hard—and that makes good horse sense. **HR**

One way to ensure more power was by switching from the stock 302 cylinder heads (left) to the '71-'72 351 Windsors. These mildly modified breathers have been fitted with big valves.

While the old cast-iron Muscle intake makes good power, Roush relied upon the 4-barrel-type Shelby aluminum marine version for a significant increase in basic horse pressure.

Among the modifications to the Holley No. R6980 carburetor was enlarging fuel feed restrictors, resetting float levels, adding splash control baffle.

You might say that this Roush-style '78-'79 302-inch Ford is box-stock. Jack estimates an engine like this can put out 270 to 280 horsepower, and if you add the "stroker" 292 truck crank, you can put a 300-horse V8 into your '79 'stang. Now that's some serious horse trading.

WIN THIS WAILER: OUR HOT ROD NATIONALS PRIZE 302 POWERPLANT

By Al Kirschenbaum

It's been nearly a decade since Motown machinery has been manufactured with muscle in mind, but the image somehow lives on. As hardcore hardware becomes harder and harder to find, ingenious rodders are left with little more than their wits to preserve the remains of an era. And now, thanks to a simulated slice of supercar life, the muscle machine look—and, in fact, a good deal of the original feel—manages to endure.

What we're referring to is replica power. "Sham" powerplants and "mock" mechanicals which are externally identical to the Sixties finest go-fast factory efforts but built with contemporary innards and modern motoring in mind. Specifically, the 302-cube "Bogus Boss" that we're investigating here is a product of the FoMoCo specialty engine works at Total Performance (40631 Irwin, Mt. Clemens, MI 48045, 313/468-FORD).

This Midwest power emporium can produce virtually anything Ford, from a Le Mans cam to a rolling Cobra classic, and owner John Vermeersch concentrates his efforts on assembling some of the finest Ford engines anywhere. John was closely involved with the factory's Muscle Parts department during the late, great, Sizzling Supercar Sixties; and he now applies all his extensive corporate skills to his own muscle-building trade at Total.

The replica engine plan, according to John, fills a natural enthusiast's need, and his shop can supply "mirage" motors of nearly any OE (original equipment) configuration. His externally identical powerplants meet a number of budgetary buildup considerations, and when following this scheme, locating the proper hardware becomes

Following the final assembly and dressing of our Bogus Boss, the entire works was trucked across town to the Engines Division of Gap Industries (13145 Wayne, Livonia, MI 48150). The 302 was then attached to Gap's Go Power dynamometer, where motor master Kenny Moe and the Total crew ran the engine through its break-in paces. When sufficient pump time was logged to assure the proper seating and sealing of all critical components, a series of power loops were run, the best of which produced a maximum output figure of 318 horsepower at 6000 rpm. The torque numbers topped out at 275 ft.-lbs. at 6000 rpm. Here Kenny twists the distributor for a timing change, under the direction of Total Performance's own big boss, John Vermeersch.

BOSS

a lot easier for anyone attempting to duplicate the efforts outlined here. High-dollar exotica, like a forged crankshaft, a four-bolt main block and rare Boss 302 cylinder heads are eliminated from the "Bogus Boss" replimotor plan as more common components are substituted throughout. Externally, the right valve covers, intake manifold and harmonic dampener set the visual facade, and internally, a solid-lifter cam or a suitably ground hydraulic will make the mimic motor sound all original too. High-flow Cleveland cylinder heads will also help make enough horsepower for competition use where a sanctioning body limits a racer to wedge-only, or non-Boss head castings. And with the exception of concours show car competitions, where a sharp judge can "bust" an entrant for the rather obviously "adapted" intake manifold, a replimotor makes a sound basis for a late-model Ford musclecar restoration.

Although our coverage of this grand illusion is fairly complete, anyone interested in a closer look at our pseudo-Ford is invited to attend our HOT ROD Nationals (see page 22), where the completed engine assembly will be on display prior to being given away as one of our many participant prizes. The masquerade muscle motor you're looking at may be a Boss Ford in name only, but the ingredients involved are all part of a basic V8 buildup brew. Follow along with us, if you can, and you'll soon fool 'em too.

Block

Available since 1964, when the 260 V8 grew to 289 inches, the 4-inch-bore small-block Ford, including the 302, is one of the easier engines to locate. And that alone makes this budget Boss buildup a timesaver as far as gathering raw materials is concerned. The original 1969 and 1970 Boss 302 was based on the same little-inch casting that we're using in our late replimotor, but the early iron packed extra-duty features like 4-bolt mains (at journal numbers 2, 3 and 4), thicker bulkheads and screw-in core plugs. These were the engineering edges needed to live at an occasional 8500. Our efforts seek the saner speeds around six grand, so as structurally sound as the Boss blocks are, they're hardly the kind of everyday boneyard fare that we'd plan a project around.

We're more concerned with this lightweight 302 "passenger car" casting that's readily available, new and used, almost everywhere. Block prep followed Total's standard deburring, bottom-tapping and cleanup plan, and the bore walls were finely finished to clear the piston skirts by a maximum of .004-inch. The final cross-hatch pattern and grade should always be determined by the individual engine's application, and John elected to finish these walls smoothly (for quick sealing on the street) with moly-filled piston rings.

Crankshaft

With the exception of the forged steel Boss 302 shaft, all production small-block Ford cranks are nodular iron castings. Our efforts were based on the common 3-inch-stroke cast shaft that's been fitted to a multitude of 302s from 1968 to the present. We're assured that a casting in good shape is well-matched to our boulevard-bound muscle.

Magnafluxing and straightness and end-play checks were performed, and factory limits were held throughout. With

the crank resting on the top halves of the main inserts and socked down at alternate caps, a dial indicator determined the runout at each journal and the overall straightness of the shaft.

Main Bearings

As with most of our buildup plans, we're following Ford's by-the-book service suggestions for refinishing a high-performance powerplant. Once we determined that our cast-iron crankshaft was in suitable shape for assembly, it was laid into the block's main saddles on a set of fully grooved TRW Clevite 77 inserts. Final assembly specs call for

main journal clearances within the .0025-.003-inch range, and that's exactly how our "simulated" screamer was screwed together.

Pistons, Rings & Connecting Rods

Open-chamber Cleveland cylinder heads dictate the use of a piston dome design compatible with the heads' unique canted valve arrangement. And since both the 302 and the 351 share a common 4-inch bore, piston swapping becomes almost a direct interchange. The OE-style TRW flat-top replacement forgings we selected required a bit of top trimming to compensate for their higher compression distance (1.647 inches, as compared to 1.600 inches for the 302), but the stock valve pockets in these Cobra Jet slugs are more than deep enough to clear our cam's planned lift.

TRW also produces 351C forgings with both high-dome and mild dish designs, but since we had to custom-tailor our 302's compression ratio through a combination of combustion chamber displacement and dome design, machining was in order, regardless of our choice. Trimming the pistons by .050-inch produced a dome-to-deck dimension of minus .013-inch, resulting in a true 9.5:1 compression ratio. Combustion chambers larger than our 76cc pieces, used in conjunction with even less deck height (there's enough meat in the TRWs to trim as much as a full tenth of an inch off their domes), could theoretically produce a low 8.4:1 ratio, while different dome/displacement combinations could conceivably kick the numbers up as high as a full 10:1.

Moly-filled TRW top rings were fitted to the 5/64-inch grooves and end-gapped at .016-inch. Similarly sized and fitted cast ductile iron second rings were installed in standard fashion, while three-piece oil ring sets completed the Cleveland assemblies. Connecting rods are standard 302 units (5.090 inches long) equipped with TRW Clevite 77 inserts fitted with .0025-.003-inch clearance at the journals. Following the factory's design plan for increased piston stability at bottom dead

center (BDC) in the short-skirt engine block, the rods' small ends remained unbushed for press-fit pin retention. Standard ⅜-inch 302 rod hardware was also maintained, while factory side-to-side rod clearance dimensions of .014-.024-inch were held throughout.

Camshaft

Our Reed Cams hydraulic bumpstick is another departure from the factory Boss plan, and although it's slightly milder than the 290-degree OE solid-lifter version, the juice grind is designed to complement our Cleveland heads' generous valves and ports. Guiding the camshaft through a small-block's decreasing diameter journals (2.08 inches at the front and 2.02 inches at the rear) is easier with added leverage from a long bolt threaded into its snout.

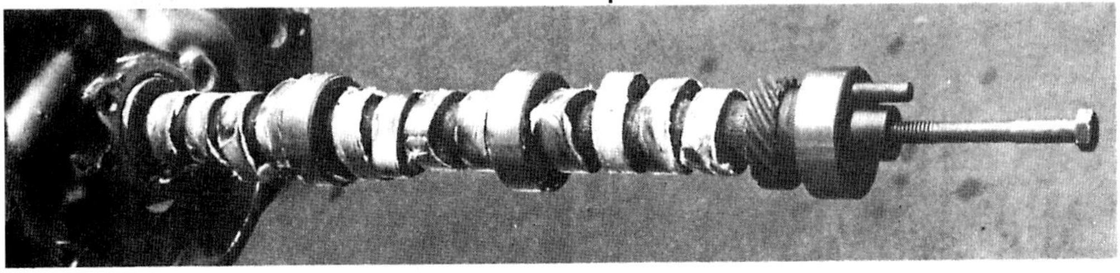

Cam retainer plates come in four different sizes. On assembly, a plate's thickness must be matched to the thickness of the center collar protruding through the back side of the cam gear and to the width of one of two available timing chains. Known combinations match gears marked "373" with a .200-inch-thick retainer, and those designated "406" mate with the .230-inch plate. Loctite is suggested on all attaching hardware in the area.

Short-Block Assembly

Factory-engine assemblies use a rope seal at the rear main for oil control, but the replacement service seal that we employed is a more performance-oriented neoprene piece. Total's engine assembly man, Kearnie Moore, leaves a short length of the seal's upper half protruding above the saddle for more effective leak prevention.

As the reciprocating assembly was screwed together, rod bolt sleeves

Continued below left.

Continued from above.

(arrows) kept the cap hardware from damaging either the finished cylinder walls or the critically smooth rod journals. Commercial "chopsticks" devices are also available for this purpose, but as you can see, simple lengths of rubber hose can serve just as well.

Various tools are available to compress piston rings into their grooves as the piston gets installed into its bore, and this simple version of the squeezer worked just fine. Kearnie used a common wooden hammer handle to tap a piston/rod assembly into its bore while he guided the bolt sleeves on both sides of the crank journal on the bottom end.

After all piston/rod assemblies were torqued down on their journals (to 19-24 ft.-lbs.), a final measurement of rod-to-rod side clearance was made. Factory specs call for .014-.024-inch clearance, and adjustments had already been made during preassembly dummying checks by grinding off one-half the amount of material needed to open up the gap from the non-chamfered faces of each pair of rods' big ends. Excessive clearance here puts added oil on the cylinder walls, where it may overcome the oil rings and contaminate the fuel/air mix. Notice the longer main bearing cap hardware designed to retain the windage tray at the intermediate journals (arrows).

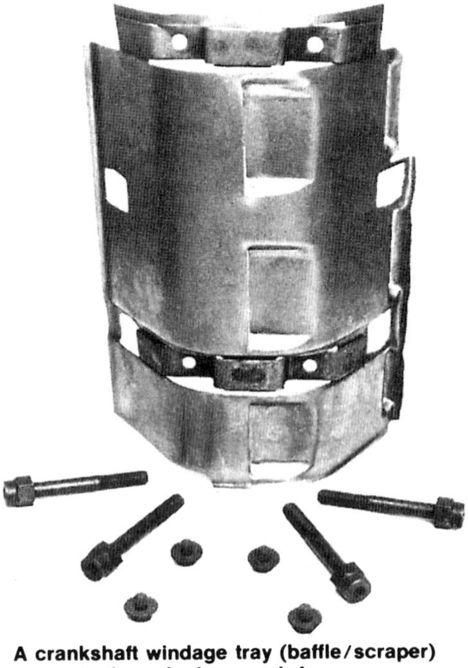

A crankshaft windage tray (baffle/scraper) prevents the spinning crank from whipping up the oil at high engine speeds. Reducing foaming and aeration in this manner will prevent the oil pump from sucking air into the system, where, if fed to a bearing, it could cause damaging metal-to-metal contact. Included in the factory baffle package are the longer main cap bolts and the necessary attaching hardware.

Cylinder Heads

Our basic replimotor plan is based on both lower component costs and the use of much more readily obtainable hardware. In the cylinder head department, this simulation scheme scores big on both points. Original equipment Boss castings are rare and therefore expensive, and replacing them with much more accessible open-chamber heads from a far more common 351C (1970-'80), 351M (1977-'80) or 400M (1977-'80) 2-barrel motor saves both time and money. There are certain considerations involved in using these replica pieces, but in the end, an assembled "Bogus Boss" costs about a grand less than a concours original. And what's more, the power produced by both 302s will be nearly equal.

The old Boss 302 heads do have slightly larger valves than the 351 Cleveland castings (see chart) and the C-pieces, in turn, have bigger valves than the popular Windsor parts. The big differences in performance potential, however, are evident in a comparison of port design and size. The combination of increased inlet and outlet areas, plus the flow-promoting 30-degree canted/staggered valve arrangement of the Cleveland castings, really accounts for the advantages of these heads. On the intake side, the Cleveland casting offers approximately 26 percent additional port area, while the numbers on the exhaust side are far more dramatic, with a whopping 100-percent-plus advantage again going the C-way.

The construction considerations involved in using open-chamber Cleveland castings on a Windsor-style 302 include the need for pistons with both the proper compression height (see Pistons) and a dome compatible with the C-heads' skewed valves. Alternator mounting may require some hardware juggling at the front of the head, but adaptation won't require much more than some choice parts-picking and simple drilling and tapping.

Mating a suitable W-style intake manifold to the Cleveland heads can be handled by simply screwing on a pair of Total's adapter plates, but some milling and/or

drilling of the heads themselves is necessary to match up the water jackets. Cleveland engines normally direct their coolant back into the block (through holes in the ends of the heads and decks) after it circulates through the heads. Cleveland intake manifolds are built to run dry. Using a 289 Windsor-style manifold with Cleveland heads requires an opening drilled or milled in the front end of the head's manifold face (arrow). These heads have been opened up beyond the required minimum water passage size. The revision works just as well with an 11/16-inch round hole at each location.

Before

After

The two passages in each head's deck surface that normally carry the coolant back into the block must be sealed off so that the fluid is redirected through the new openings in the intake manifold faces. Since these portions of the heads overlap the block, a number of methods can be used to close the openings. Welding the four holes shut will work, as will a pair of epoxied shim stock blockoffs. The plugging procedure illustrated here is slightly more complicated, but it's probably the most effective plan of all. John used a vertical mill to cut a 1-inch round lip into each square passage, after which standard flat Dorman plugs were pressed and sealed into place.

The stock stamped-steel Cleveland rocker arms and fulcrums are more than up to handling our avenue-inspired application. Cam lifts in excess of .550-inch will, however, require screw-in pivot studs, and the slots in the rockers will have to be carefully measured and inspected for possible interference problems at the higher lifts. If longer pushrods are

properly employed (with a higher-lift cam), it's possible to adjust the stock rocker geometry to suit the standard slots.

VALVE SIZES
(inches)

	289 302	Boss 302	351C	351W* 351M** 400M**
INTAKE	1.75	2.20	2.04	1.84
EXHAUST	1.45	1.70	1.65	1.54

*Ford GT40 or Chevrolet pieces fitted to a 351W head can yield a maximum 1.875-inch intake valve and a 1.625-inch exhaust valve.

** Slight variations in specific model runs produced identical cylinder head castings with slightly smaller valve sizes.

Intake Manifold Adapters

The most advantageous OE-style intake manifold available for this unique application is an easily located piece on the used parts market, but a number of aftermarket designs can also fill the bill. The popular small-port 289 inlet casting we selected can be found on any number of early Mustangs, Comets and Fairlanes fitted with the standard 4-barrel V8, rather than with the rarer high-performance version of the 289 (see August '79 HOT ROD for 289 identification tips). Visually, the 289 dual-plane manifold closely resembles the OE Boss casting, and internally, the port match with the Cleveland heads is close enough to need no reworking. (A Boss manifold, on the other hand, produces a radical mismatch with all heads except the original Boss and the 351C 4-barrel heads.)

The only problem arising in this application, mating the manifold to the

heads, requires these adapter plates. Designed to rearrange the attaching bolt pattern rather than to align the ports, these fully machined 3/16-inch-thick alloy plates are available from Total for $89.95 per set, including gaskets. Two new holes must be drilled and tapped (5/16 coarse thread) in each head's manifold face to retain the plates, and the pair of existing intermediate holes in the heads must be plugged (with interference-fit threaded rod) and redrilled and tapped at the correct 45-degree angle to suit the manifold's hold-down holes. The round opening at the end of the plates can be used as a guide for the new water passage required in each end of the heads.

Oiling

For serious street service, a stock steel oil pan was extended outboard for added lube capacity (8½ quarts) without sacrificing ground clearance. Internal "accel" and "decel" baffles are employed to control "slosh."

Production small-block oil pumps (with the exception, again, of the Boss 302 unit) are cast-aluminum items. Our replacement from TRW is also a cast-iron part with a longer rotor and a deeper body for increased volume.

As seen from below, where it attaches to the block near the number one bearing cap, the TRW pump's thicker bottom cover adds stiffness to the assembly. The "sucker" pipe and screen are the larger-capacity (⅝-inch inside diameter as compared to ½-inch standard) Boss 302 pieces. HR

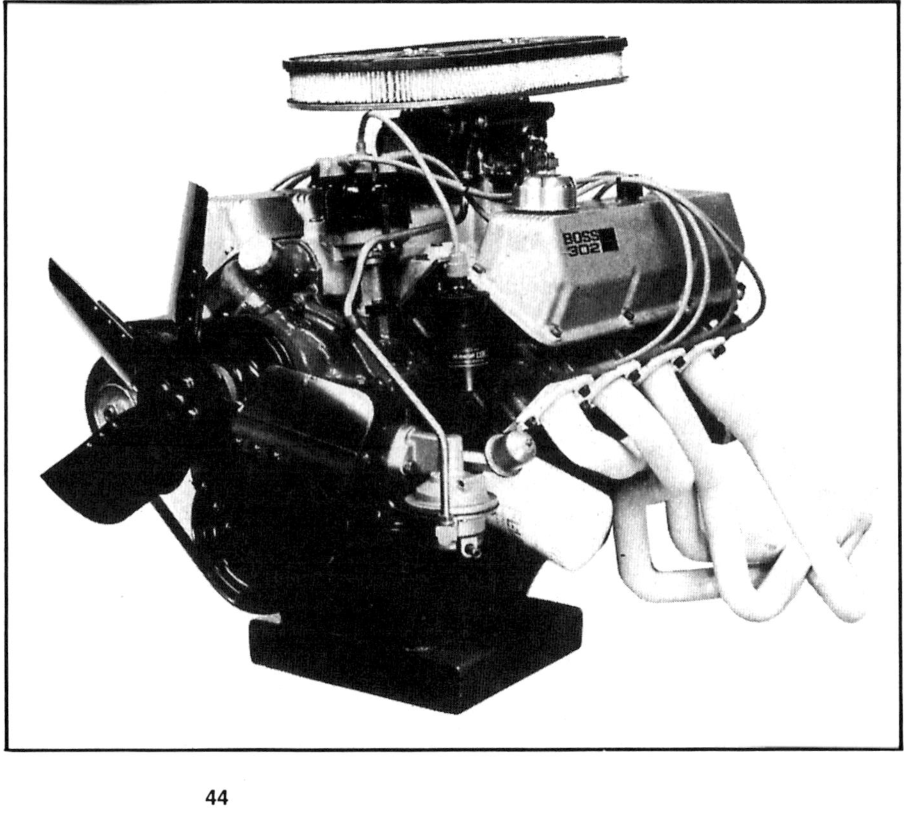

Affordable performance

MODIFYING THE FORD 351 CLEVELAND TO DELIVER AN HONEST 600 HP AT 9500 RPM

By Marlan Davis

The Ford 351 Cleveland can produce more horsepower per cubic inch with carburetors on gasoline than any other American motor. The canted valve head design allows huge diameter valves for its displacement size, and with proper head work, especially on the exhaust side, the Cleveland has proven to be virtually unbeatable—when not forced by various sanctioning bodies to carry extra weight. But since it doesn't come stamped with a "bow tie," parts availability tends to be somewhat limited, and those parts that can be found are usually more expensive than other "Brand X" pieces.

One man who has persevered and overcome this problem is Mark Degroff, general manager at Valley Head Service, 19340 Londelius St., Northridge, California 91324, (213) 993-7000. When he's not machining heads or building engines for customers, Mark's out racing a C/Modified Production '65 Mustang (see HOT ROD, August '79) with partner Doug Wright.

For the 3400-pound car to run consistent 10 teens at 135 mph (Mark has gone as quick as 10.03) in the 9.7-pound/cubic-inch C/MP class, requires a 357-inch (.030-over 351) Cleveland that puts out an honest 600 hp at 9500 rpm, that's backed by a 3.25 low gear Doug Nash 5-speed and 6.50-geared Dana rearend. That kind of rpm requires an engine nearly as radical as a Pro Stock motor, yet Mark feels his combination could be duplicated for around $6000, assuming the builder does all the work

PHOTOGRAPHY: BRUCE CALDWELL & MARLAN DAVIS

Affordable performance

(except for major machining and headwork). However, a radical high-output motor such as this one does require frequent major maintenance: The ring seal goes away after about 25 runs and a valve job is necessary after 50. After every race, the engine is torn down and inspected.

BLOCK PREP

Mark begins with a used 4-bolt-main Cleveland block, which came stock on the '71 Boss 351. A few other passenger blocks, including some from Mercury stationwagons, have also been known to turn up with 4-bolt-mains. Two-bolt-main blocks have enough meat to add the extra bolt holes, but aftermarket caps would have to be used since Ford doesn't sell new caps without an engine. Used blocks are preferable to new ones because their cylinder walls have already shifted as much as they're going to.

Only three machining operations are required to prepare the block for racing—align boring, cylinder wall honing and deck surfacing. According to Mark, "We don't strive for any particular deck height on the block because we individually adjust the piston by machining it to achieve a nominal zero deck with the head gasket installed." Normally only a .005-inch surfacing cut is taken, and the block is left as tall as possible.

Moroso restrictor kit is installed to prevent oil starving mains. Galleys feeding into No. 5 main saddle *(circled)* are most important. Middle hole goes to cam and is tapped to receive 5/16 coarse thread restrictor. Left-hand hole is left alone as it feeds oil from pump via passenger-side lifter galley to this bearing. Hole on right sends oil from bearing to driver-side lifter galley; it's restricted with ⅜ coarse plug. Small galleys in saddle Nos. 2-4 that feed oil to cam are tapped for ¼-inch-20 .040 restrictor plugs *(arrows)*. No. 1 hole is left alone; since main oil galley passage from filter and cam passage criss-cross underneath main web, plugging it would only restrict oil to bearing.

Standard Moroso small-block Chevy windage screen is modified for use on Cleveland by bending to semi-circular configuration and trimming to fit crank cavity, as shown. Original modified screen ended at No. 4 saddle, but rod failure has since sliced it off. Block is tapped on outside edge for ¼-inch-20 bolts that retain homemade sheetmetal brackets.

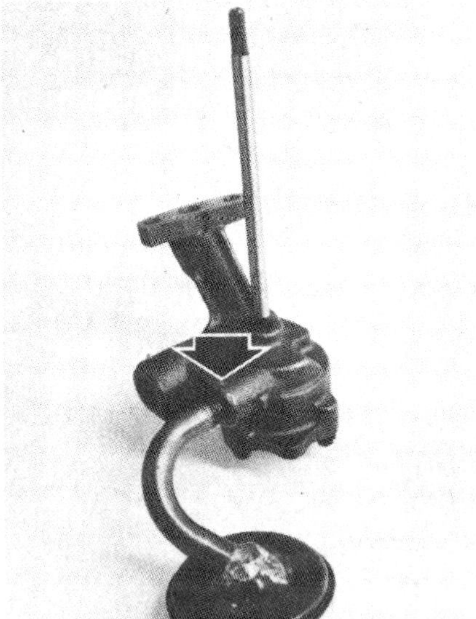

Standard TRW high-volume oil pump is used with normal passenger-car pickup brazed onto screen as insurance against factory weld breaking. Other end is securely fastened into pump body with ⅛-inch pin replacing the factory screw-in method. Pin is peened on both ends to achieve tight seal and prevent minute vacuum leaks which can aerate oil as it's sucked through blades. Ears are welded onto screen so in event of oil pan failure, pump-to-pan clearance will still be retained and oil starvation prevented. Pump driveshaft is bulletproof steel piece from Pro-Stock Engineering, Inc., 16102 Orange Ave., Paramount, CA 90723.

Cylinder overbores have ranged from as little as .020 to as large as .060 (his current 351 is .030 over), and are done using a torque plate to simulate cylinder head bolt distortion. Cylinder walls are finished as smooth as possible on the Sunnen CK-10 honing machine. Mark starts with a 625 stone (approximately 400-grit), progressing to an 820 (about 600-grit). Then a special ultra-fine C3003 800-grit stone is used for the final .0002-inch.

OILING

Unlike some other Cleveland engine builders, Mark does not perform any radical oiling system mods. A simple Moroso oil restrictor kit is used to limit oil flow to the driver-side lifter galley and cam journals *(see photo)*. All you have to do is tap the holes and put in the Allen screws, which come predrilled to the correct restrictor size.

Other key oil system ingredients include a TRW high-volume oil pump, TRW fully grooved main bearings and Crane roller lifters featuring a very small waist that allows oil to flow more easily past it. This feature is desirable since the Cleveland feeds the rear main bearing through the passenger-side lifter galley.

RECIPROCATING ASSEMBLY

Ford never offered a forged crank for the Cleveland, but Mark has found the stock cast-iron unit, as reworked by Velasco, to be more than adequate. The crank is cross-drilled for 360-degree oiling and Tufftrided. To

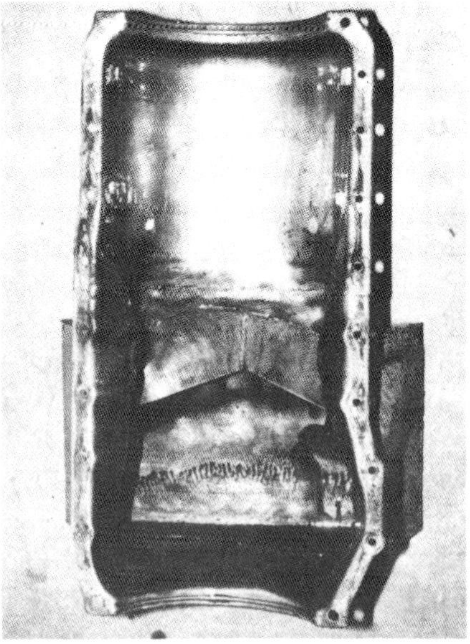

Instead of being deepened, stock 351 pan was widened to increase capacity to nine quarts (with filter) from stock five. This allows stock length pickup to be used and helps ground clearance. (As seen from this pan's slightly bulged-in sides caused by post-wheelstand letdown you can't have enough!)

better cut through the oil at high rpm, the counterweights are streamlined or "knife-edged." On cars that see a broad rpm range, such as Mark's 5-speed Modified Mustang, this mod is reportedly worth as much as 30 hp and 2 mph on the top end. Cars that experience little rpm fluctuation—such as a Powerglide-equipped Econo dragster—do not seem to benefit all that much.

The crank's rod throws are cut down from the stock 2.311-inch to the Chevy small-block's 2.1-inch diameter. Bearing speed is thereby slowed, which translates to more rpm capability and less chance of spinning a rod bearing. What's more, it's now easy to stroke the crank as much as .400-inch up or down from stock, since the journals are being cut anyway. Mark is currently running a longer-than-stock 6.1-inch aluminum Super Rod, which he prefers because it's the lightest aluminum connecting rod on the market, yet is still strong enough to last about 80 runs. With the stock stroke crank in his present engine, the "ideal" 1.8:1 rod/stroke ratio is achieved. Although a shorter stroke would allow more rpm, the heads seem to feed best on a motor displacing around 350 inches.

For pistons Mark uses a gas-ported Venolia Pro Stock racing unit run with .007 cylinder wall clearance. The gas porting promotes better high rpm sealing in conjunction with the special Sealed Power ring set, consisting of a .043-inch plasma moly top and 1/16 tapered face cast-iron second (gapped at .014 and .012, respectively), along with a special SS50U ⅛ low-tension chrome-faced stainless-steel oil ring. Mark also feels the Venolia dome configuration offers good flame propagation, while the pistons' ground-round skirt is easy on the cylinder walls. As it's among the thinnest of American engines, the Cleveland's walls tend to crack when exposed to the more common slipper-type skirt.

Crank is internally balanced with Mallory heavy metal in front *(shown, see arrows)* and rear counterweights, reducing load and strain on No. 1 bearing caused from supporting very heavy counterweighted dampener located two or three inches in front of it. To aid high-rpm stability, engine is actually overbalanced 52 percent. Counterweights are knife-edged to cut through oil better at high rpm. Studs are used on inside of 4-bolt-main bearing caps since stock Ford bolts aren't strong enough to withstand 100 ft./lbs. of torque called for in racing use. They're more than adequate for outside holes, where only 45 ft./lbs. is needed. Rod and main bearings run .0025-.003-inch clearance. Thrust clearance is .005-.008-inch.

Each piston is recontoured by hand all the way around so it fits the combustion chamber as closely as possible when the motor is first assembled. As temperature and rpm increase, the dome actually gets so close to the cylinder head in certain areas that the carbon buildup is brushed off (see photo). As Mark says, "It's one way of verifying the piston is as close as it can possibly get!" So far, that comes out to .050 piston-to-valve clearance on the intakes and .080 on the exhaust. More clearance is required on the exhaust side because the piston is chasing the closing valve and would hit it if the valve momentarily floated.

As the life of the engine progresses and the connecting rods continue to stretch and take a set, a file is periodically taken to the piston's quench area where it's hitting hardest. Then .003-.004-inch of stock is taken off, enabling the motor to get another 15-20 runs in before the procedure has to be repeated.

A relatively large rod side clearance of .050-inch is run to avoid any restriction on total oil flow through the bottom end. Although this always keeps the bearings cool, such large clearances may throw additional oil on the cylinder walls. However, the gas-ported pistons and special ring

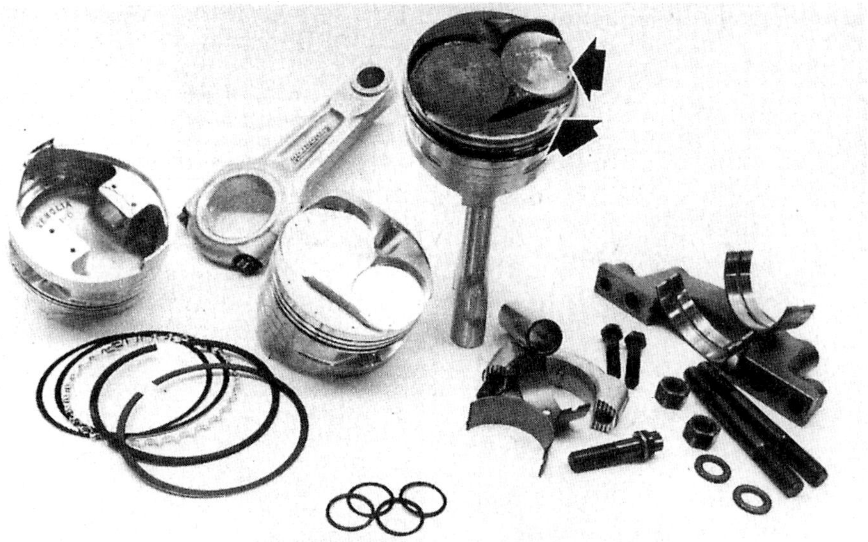

Venolia pistons are retained on Super Rods by standard Venolia 66-gram, 2.5-inch-diameter non-tapered, thinwall wrist pins held in place by dual Spirolocks on each end. Special Sealed Power ring set designed for gas-ported pistons is used. Domes are smoothed and recontoured by hand to fit chamber as closely as possible, giving aproximate 14:1 compression ratio. Fit is so tight that carbon is actually brushed off by chamber roof (note light spots on dark carbon-covered used piston). Fully grooved TRW MS-3010B main bearings fit in stock Ford caps, aid oiling. Caps are retained by stock Ford ⅜-inch bolts in outside holes and Automotive Racing Products ½-inch chrome moly screw-in studs with hardened nuts and washers. Since rod journals have been cut down to 2.1-inch Chevy size, TRW 350 Chevy rod bearings (part No. CB-824P), predrilled for retaining dowel pin, are used.

Since engine is internally balanced, factory counterweight on stock balancer *(bottom)* is machined out. Dampener outer portion has reputation for separating from rubber layer, so holes are drilled and tapped opposite each other *(arrows)* for bolts that act like dowel pins to keep outer rim indexed while allowing rubber to flex and retain its dampening effect.

Affordable performance

combo seem well up to handling it.

One other trick Mark performs on the bottom end is to replace the stock rope-type rear main seals with Fel-Pro BS40042 rubber pieces. This is done because the stock seals put excessive drag on the crank.

CAMSHAFT AND VALVETRAIN

The cam used is a Crower roller grind (part No. S-0291) ground on

Bulletproof pieces are required to handle .700 lift cam at over 9500 rpm. Crane No. 52718 roller lifters push cut-down TRW chrome moly big-block Chevy ⅜-inch pushrods. Then 1.625 o.d. Crower Vascojet springs (part No. 68555-1), capped with titanium retainers held in place by 10-degree split locks (part No. 86068), are adjusted by hardened stainless-steel shims to installed height of 1.950 inches, giving 200 pounds of seat pressure (750 open). Standard 1.73:1 Norris big-block Chevy rocker arms are used on Automotive Racing Products chrome moly rocker arm studs (part No. RRS1-7/16). Standard size Del West 11/32-inch stem titanium valves (2.19 intake/1.70 exhaust) are protected from wear caused by rocker arms with Competition Cams lash cap. Guideplates are modified Chevy rat parts.

Stock Chevy ⅜-inch pushrod guideplate (part No. 3860038) *(top)* is cut in half and bent slightly on ends for use in Mark's Cleveland; necessary because distance between 351's stud centerline and pushrod is less than on Chevy. It's practically a bolt-on operation—just install on head, bend, index for proper relationship to pushrod and torque down stud.

109-degree lobe centers with 71 degrees of overlap and .785/.787 lift and 328/332-degree duration on the intakes and exhaust, respectively. Mark gave Crower all the pertinent information on the engine, drivetrain, chassis and intended application, along with complete data on head flow rates at varying valve lifts, and they did the rest. Says Mark, "It's the best cam I've ever run." If he could, he'd use a rev kit with it, except that no one offers such a setup for the Cleveland.

A variety of different manufacturers parts are used to complete the valvetrain *(see photo).* Many are modified from parts originally developed for the big-block Chevy, which has similarly designed cylinder heads with canted valves and independent stud-mounted rocker arms. Even the two engines' valvestem heights are within .025 of each other; the face diameter of the intake valves (2.19 inches) is exacly the same. Mark has found a

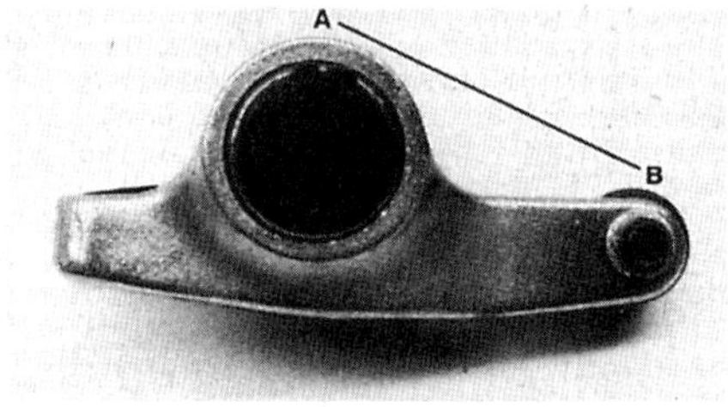

Overall distance on this Norris big-block Chevy rocker arm between trunion (A) and roller (B) is .020-inch longer than on Ford piece. With radical rocker arm movement generated by high-lift cam, added length means roller will be closer to center of valve stem, resulting in less wear.

Pro-Stock Engineering's ProDex timing chain and indexed gearset permits easy changes of cam timing. Cam is normally run straight up, but at times is retarded as much as 2 degrees to ensure proper relationship to piston if dome or rod stroke is slightly off. Occasionally cam is advanced 1 degree to compensate for chain stretch. Object is to "run cam as close as possible to what Crower recommends."

part ostensibly made for a "big-block Chevy" is usually cheaper than one "special ordered" for a "351 Cleveland."

Since the two motors aren't *exactly* the same, some mods are required when using certain Chevy parts. For example, the TRW big-block Chevy ⅜-inch-diameter pushrods are almost ¾-inch too long (due to the rat's higher deck), so the pressed-in tip on the guideplate end is removed in a lathe, the pushrod is shortened, and the tip is reinstalled under one-ton pressure. This will provide an inexpensive—yet effective—Cleveland-style unit. Guideplates can also be modified from Chevy parts, as shown in the photos.

CYLINDER HEADS & INDUCTION

The heads Mark likes to use are (in order of desirability) the Boss 302, Boss 351 and 351-4V closed chamber. Boss 302 heads can be adapted

Closed chamber head *(left)* is preferred over open chamber version since it's easier to get compression up.

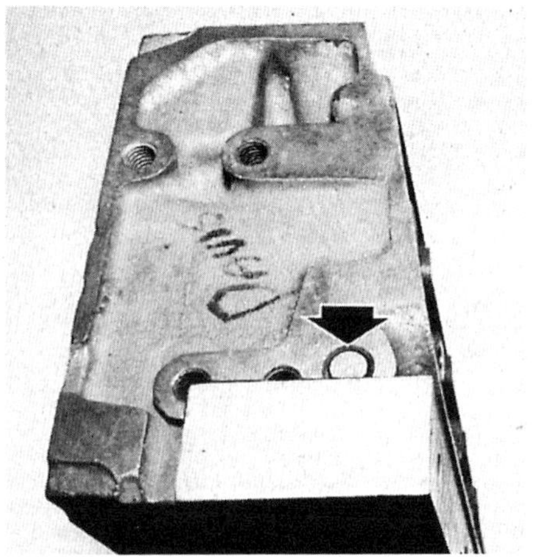

Most desired head is '69-'70 Boss 302 because it can achieve better exhaust port layout. Unlike Cleveland head, it has a cast-in smog device hole *(arrow).* A 7/16 hardened steel shaft can be inserted in this hole, allowing porter to cut higher into exhaust port roof than on Cleveland. Result is a better flowing port.

Affordable performance

to fit the Cleveland with minor mods, and their design allows more work to be done in the exhaust port area than in the other two. These heads, as well as those of the Boss 351, are machined for screw-in rocker arm studs and pushrod guideplates, unlike the 351-4V heads (although they can be modified to accept them). Closed chamber 4V heads are preferred over late-model, open chamber smog heads because it's easier to get the desired 14:1 compression with the former. Of course, 2V heads of any year are totally undesirable for drag racing.

Whichever head is used should be surfaced to achieve a 55cc combustion chamber. Typically, this requires a .060 cut; angle-milling is unnecessary since there's more than enough meat on the heads.

All heads receive a full competition port and polish. The ends of the stock exhaust ports are cut off and replaced with a raised aluminum plate, a mod which is relatively straightforward nowadays. The intake port is thoroughly hogged out and streamlined around the valve. Mark cuts as high as he can into the port roof, trying to straighten it all the way from the valve guide as far up the Edelbrock tunnel ram as possible. The valves get a special Valley Head Service valve job and are installed in special K-Line silicone-bronze .015-inch-diameter thinwall valve guides, which are sealed with Perfect Circle Teflon pieces that fit inside the Crower valvespring.

Finally, the heads are O-ringed to complement a special McCord .050-inch compressed thickness composition head gasket (part No. 6850). With the O-ringing and this gasket, the heads need only 80 ft./lbs. of torque to seal them (compared to the recommended stock figure of 100-125 pounds). Mark feels that lower bolt torque causes less distortion of the heads, leading to much better valve sealing.

The motor is topped off by two Holley 650 double-pumpers (Model 4150, List 0-4777). Mark prefers these carburetors over the huge 1050/1150-cfm 4500 Dominators because the larger carb doesn't seem to respond as well as the 650 to the sudden drastic rpm changes experienced in a manually shifted car. Since the Edelbrock UR-19 tunnel ram is designed for the 4500, adapter plates had to be fabricated. The smaller size 4150 carb provides additional distributor clearance, so the carbs can be moved further forward than the 4500 on the intake for optimum positioning directly over the manifold plenum, thus giving fuel a straighter shot into the ports.

The carbs are run surprisingly stock: Naturally, jetting varies, depending on track and atmospheric conditions. The jets are staggered to compensate for rich and lean cylinders. No power valves are used. The air horn is milled off for better air flow—since the choke's been done away with it's redundant anyway. To prevent loading up at idle, or flooding the engine with an overly rich charge when mashing the brakes after a burnout, the idle metering circuit is restricted with .017-inch-diameter wire.

The combination, when properly maintained, consistently provides Mark with afFORDable performance. Why not try some of his better ideas? **HR**

Cleveland head *(front)* doesn't have rear water crossover passage like Boss 302 does. To use Boss 302 heads on Cleveland block, this water passage must be blocked off.

OPERATING AND TUNING DATA

Fuel:	Daeco 105-octane, ½-inch fuel line, Holley electric pump
Fuel Pressure:	8 pounds with dual pressure regulators
Oil:	Valvoline 20W-50 racing.
Oil Filter:	Fram
Oil Pressure:	60 psi (average)
Thermostat:	None (block run through quarter dry)
Normal Operating Temperature:	180 degrees off line; boils over in lights
Ignition Timing:	40 degrees total, fully advanced at idle
Spark Plugs:	Autolite AF-701 racing
Plug Gap:	.040-inch
Valve Lash:	Intake—.026-inch (hot), .024-inch (cold) Exhaust—.028-inch (hot), .026 (cold)
Carb Jetting:	Stagger-jetted, 74 average
Power Valves:	None
Accelerator Pump Squirter Diameter:	.035-inch
Headers:	Home-built using Hooker kit: 2¼-inch-diameter primaries, 24 inches long; 4-inch- diameter collector, 10 inches long, dictated by chassis considerations

Here is what completed engine looks like, topped by 650 Holley double-pumpers minus choke and airhorn. Original photo-electric-triggered Accel BEI lights spark. Water pump is rare Ford aluminum muscle part piece (part No. DOZX-8505-A). No water actually circulates through block during run; pump is electrically actuated by Mr. Gasket water pump drive kit (part No. 4333) and circulates cool water from radiator through block on return road. Need to add water manually is eliminated, except in event of boilover.

GET NEW TFS/KAUFMANN ALUMINUM 302/351W HEADS AND . . .

CROWN YOUR WINDSOR

By Marlan Davis

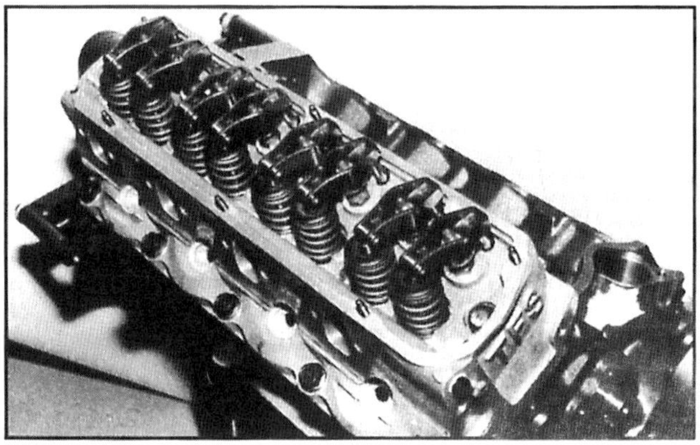

New aluminum heads come machined for stock 11/32-inch stem valves and 1.550-inch O.D. valve springs. For oversize springs, the pockets can be deepened .050 inch and/or widened to 1.624-inch O.D. Rail-type rockers work well, eliminating the need for guideplates.

Pity the poor Ford Windsor family! They always wanted to get *ahead* but until recently were stymied by their Boss, the Clevelands. All the high-perf factory and aftermarket R&D went into canted-valve Cleveland head development. While that may have been fine for bucks-up professional racers, it didn't help the vast majority of small-block Ford owners, who hailed from the Windsor side of the Ford small-block family tree. For them, installing a set of Cleveland heads was an outrageously expensive proposition, usually requiring a new cam, valvetrain, pistons, and rods—almost an entirely new motor! What street rodders, street machiners, and assorted weekend warriors really needed was a state-of-the-art high-perf head that would bolt in place of existing Windsor heads with minimal modifications required.

Enter Trick Flow Specialties (TFS), the people who brought us the high-tech 429/460 race heads, and Kaufmann Products, pioneers in hopping up late-model, high-tech Ford V8 production hardware. They have collaborated in the production of a new, aluminum, Windsor head that features modern swirl technology and improved intake- and exhaust-port configuration, yet it is still compatible with most production Windsor blocks, valvetrain components, and smog devices. The new head has flow characteristics superior to any non-canted-valve small-block Ford head built to date and is 50-percent lighter than its cast-iron predecessor. (That's a quick 45 pounds off your front end.) Sound intriguing? Let's take a closer look.

INTAKE PORTS

Compared to stock heads, the port is longer, its roof is raised, and the short-turn radius is improved. TFS has obtained swirl-related patents on certain design features of their big-block head—features that they have also incorporated into the new Windsor head. Understandably, they're reluctant to go into specifics. Suffice it to say that the new design involves port and combustion-chamber shape and geometry, relating to the "controlled helical pitch" of the "angle of the air as it approaches the centerline of the valve stem." Keeping airflow velocity high is also said to be important.

COMBUSTION CHAMBER

As is typical with modern wedge-head chamber design, the spark-plug wall is laid back with a shallower angle, and corresponding material is added to the opposite side between the valves to maintain chamber volume and promote the desired swirl effect. The exhaust-valve location within the chamber has been slightly relocated to aid the swirl and to provide room for oversize valves and the mandated valve-seat inserts. TFS says the change in exhaust valve

In comparison to the stock intake port (left), the TFS port is longer with a straightened roof and improved short-turn radius. This results in an intake-port configuration that Kaufmann says "is more efficient than the best possible porting of any factory cast-iron head."

TFS redesigned the combustion chamber (right) to promote swirl, slightly relocating the exhaust valve in the process. Nominal chamber volume is about 58 cc, which is about equal to most 302 heads. The seats come roughed in to accept 1.94-inch intake/1.60-inch exhaust valves, but there's sufficient clearance for 2.080-inch intake/1.650-inch exhaust valves. Typical EFI/computerized vehicles use 1.78-inch intake/1.45-inch exhaust valves.

This early TFS/Kaufmann prototype casting (top) compared to a 302 iron head, (bottom) had a slightly raised intake port entry face. Testing showed that virtually the same performance is obtainable with a stock entry face location, which also ensures stock intake manifold interchange. Restrictive pushrod bulge (a) is gone. Although the aluminum head has a 351W-style water passage and an extra intake manifold bolt hole (b), seal on the 302 is unaffected, since the hole and passage are still within the square 302 gasket's envelope.

location "allows real airflow and wetflow gains around the intake valve." Kaufmann adds, "The modifications to the short-turn and chamber angle have created a much cleaner swirl. The stagger of the intake/exhaust valve relationship has helped provide a better lean burn."

Despite the chamber revisions, no special pistons are required unless very high lift cams (over .550-inch lift) or ultra-high compression pistons are used. TRW 12:1 pistons work fine for competition, but a piston dome specifically configured around the new chamber shape offers potential power and compression-ratio gains.

EXHAUST PORTS

Current production Ford V8 exhaust-port design dates from the '60s, when Henry's unibody cars all had high shock towers. The tight clearances imposed by the boxy structures forced compromises in port design and layout that are still reflected even in the factory's modern offerings. Hence, major improvements in overall cylinder-head efficiency are obtainable by designing an exhaust port unfettered by obsolete design considerations. The new head features a revamped exhaust port with a longer, smoothly swept radius that exits ¾-inch higher than stock. Kaufmann already has redesigned headers for '79-up Fox chassis (Mustang, Capri, T-Bird, Fairmont, etc.).

VALVETRAIN

Ford Windsor heads have used several different valvetrain configurations through the years. The new TFS/Kaufmann heads are designed to utilize screw-in 7/16-inch rocker-arm studs (Ford Motorsport M-6527-B311 or equivalent) and corresponding stud-mounted rocker arms and fulcrums; they won't work with the '77-up pedestal-type rocker-arm setup.

Because of the slightly relocated valves, existing pushrod guideplates may require modification to work in the new head. For this reason, TFS/Kauffmann recommends the use of high-strength, investment-cast "rail" rockers (Competition Cams' part No. 1431 or equivalent), which work very satisfactorily without guideplates. Despite the slightly changed valvetrain geometry, Kaufmann says stock-length pushrods (Kaufmann Products part No. 2238 or equivalent) work well when used with

GO WITH THE FLOW

The new TFS heads are superior to any cast-iron Windsor head virtually right out of the box. "Stock" TFS/Kaufmann port volume is about 210cc on the intake, and the exhaust displaces about 80cc. At first glance, this appears much larger than the typical "E5AE" 1985 302 HO head (125cc intake, 40cc exhaust) or '69-'70 351W (142cc intake, 45cc exhaust), but roughly 20cc of each port's additional volume can be attributed to the TFS head's longer port centerlines. Porting enhances the alloy head's performance even further, but it should be performed by a shop or individual familiar with the latest in high-swirl technology. Bigger isn't necessarily better. Velocity, wet flow, and specific port shape is where it's at. Specific porting will depend a great deal on what intake/header combination is being planned. For example, if EFI (Electronic Fuel Injection) is used, the limited performance potential of the late computer system only requires that the intake manifold and head be port-matched. If the recommended Kaufmann 1¾-inch primary tube headers are used with the EFI system, the exhausts should be ported in favor of the roofs and sides. If smaller headers are used, very little port work is required. The chambers should have the seat transition blended into the valve bowl area, and the exhaust side of the chamber can be polished.

Head porters have just begun to investigate the new head's potential. Chris Kaufmann compared an '87 Mustang iron head to the new aluminum head. The heavily massaged iron head was hogged out to the maximum extent possible without adversely affecting street durability and was fitted with oversize 1.90-intake/1.60 exhaust valves. (Kaufmann calls it his "Swirl head, Stage 3" configuration.) By contrast, the aluminum head was run virtually untouched, with only a minor cleanup and three-angle valve job. It was fitted with the same, slightly smaller than standard, 1.90/1.60-inch valves to ensure a fair comparison. (All flow figures are in cfm, based on a 4-inch orifice at 25 inches of water pressure using a Superflow bench.) The new aluminum heads outflowed the cast-iron design from .200-inch lift on up, with the disparity steadily increasing at the higher lift numbers. At .600-inch lift, the aluminum head's intake port had a 21-percent flow advantage, while at 31 percent the exhaust side's disparity was even greater!

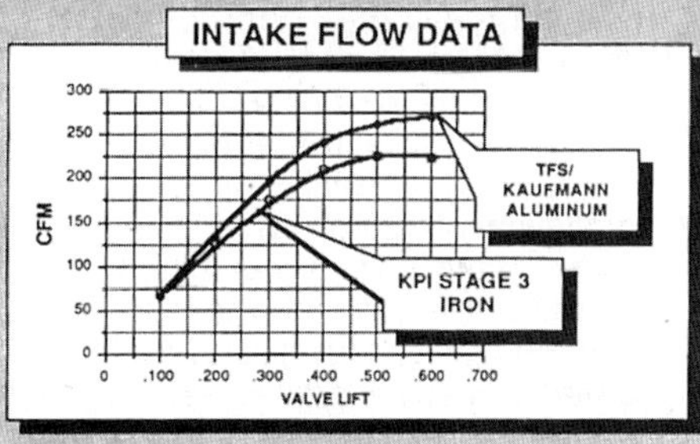

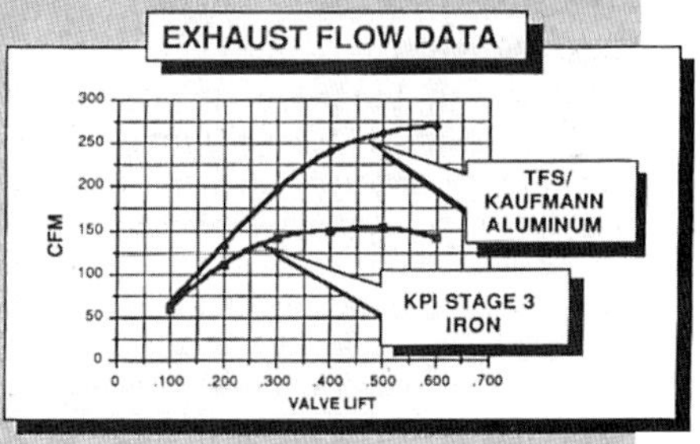

Exhaust ports on the TFS head (right) exit about ¾-inch higher than stock. The new, much straighter port offers dramatic flow improvements over the sharply curved stocker, but custom headers may be needed with some chassis combos. The raised exhaust ports retain the narrow, stock mounting-bolt spacing, so adapter plates are required to install large-tube race headers. Stock valve covers fit since the cover rail remains at its stock height and location.

standard-length valves. With .100-inch longer valves use Kaufmann part No. 2230 6.905-inch OL pushrods; with 302 hydraulic roller cams, Kaufmann part No. 2234; with 351W engines, Kaufmann part No. 2237.

INTERCHANGE

The heads are machined for ½-inch o.d. head bolts, the same as a 351W. This requires that 289/302 blocks be drilled and retapped oversize, but it dramatically improves sealing and head-gasket retention. (Fel-Pro's part No. 1011-1 heavy-duty gasket is recommended.) Other than the head-bolt differences, the head will interchange with all 289/302W/351W engines, provided (as noted above) that post-'77 valvetrain components are replaced with adjustable components. Ample deck thickness is provided to maintain adequate deck strength with straight or angle mill cuts of up to .100 inch, and all late-model accessory-mount bolt bosses are included. (Because late-model power-steering pump brackets use a larger 7/16-inch o.d. mounting bolt, earlier brackets must be drilled out oversize.)

THE FUTURE

As we go to press, the heads are available from TFS or Kaufmann. They can also be ordered with or without the internal smog pump tubes machined. In the configuration described in this article, the heads go for $995 per unassembled pair with the 1.94/1.60-inch valve job roughed in. The customer can request custom modifications, such as larger valve seats or machining for oversize valve springs, for an additional fee.

While intended for high-performance street use, the new heads could also find a home on the short-track, where the performance of the pure in-line wedge configuration is superior to the low-end and off-the-corner performance of the canted-valve Cleveland design. Production of an all-out race version with raised intake ports and further revised valve locations is a possibility. In the meantime, preliminary tests of virtually "stock" new heads have shown significant improvements when compared to heavily ported iron heads (see sidebars). If used with other properly matched engine components, your Windsor can now be King—which is better than being Boss! **HR**

DYNO-MITE

In an attempt to find out how a set of the new aluminum heads would perform in comparison to their cast-iron predecessors on an otherwise identical engine, Kaufmann built a blueprinted Advanced Engineering West short-block and installed it on the state-of-the-art computerized Superflow engine dyno located at Nitrous Oxide Systems in Cypress, California. Kaufmann held the compression at 8.8:1, simply improving the durability of the engine with B&B main cap studs and ½-inch head bolts. The engine was then subjected to over 54 dyno "step tests" from 2500 to 6500 rpm.

The iron and alloy heads were modified as described for the flow-bench tests. Two separate tests were made first using the Ford computerized EFI system with a Kaufmann-modified intake manifold and secondly with a 750 cfm Holley double-pumper with a "four-hole" idle system (0-8156) atop a 1-inch spacer and a non-port-matched Edelbrock Torker II intake manifold (part No. 5021). *All smog equipment except the catalytic converter was hooked up during the tests.* Both combos used Trick 91.5 octane unleaded gas, a custom-ground Kaufmann solid-lifter cam (.528-inch intake/.541-inch exhaust lift, 220 degrees intake/235 degrees exhaust duration at .050 lift, 112 degrees lobe separation) and headers with 1¾x31-inch primary tubes and 3½x12-inch collectors. The aluminum heads' superior heat-rejection characteristics allowed 41 degrees of timing; only 35 to 36 degrees of advance could be run with the cast-iron heads.

On the fuel-injected engine, the iron heads made more midrange torque, but the aluminum heads delivered more top-end horsepower. Probably low-end flow velocity is less on the aluminum heads due to their larger ports. The factory EFI manifold is also fairly restrictive. With none of the fuel-injection system's restrictions, the Holley double-pumper really allowed the new heads to breathe, and they outpulled the cast-iron heads above 3000 rpm. Although the torque and power peaks remained virtually unchanged, the aluminum heads produced about seven more lbs.-ft. of torque and 23 more horsepower at these points than the iron heads (for a max of 327.9 lbs.-ft. at 4750 rpm and 361.7 hp at 6250rpm). Valve float occurred at 6300 rpm.

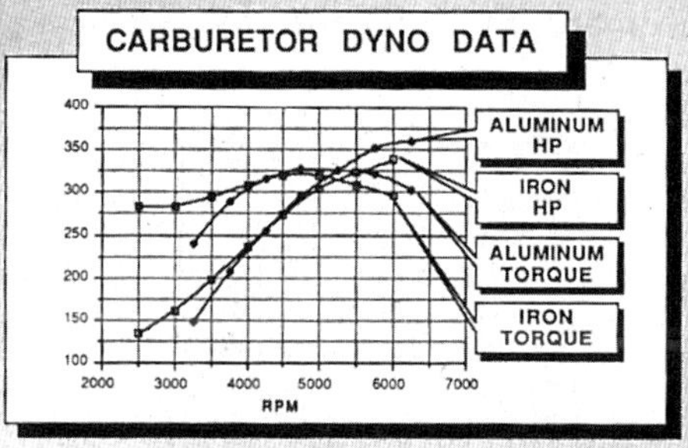

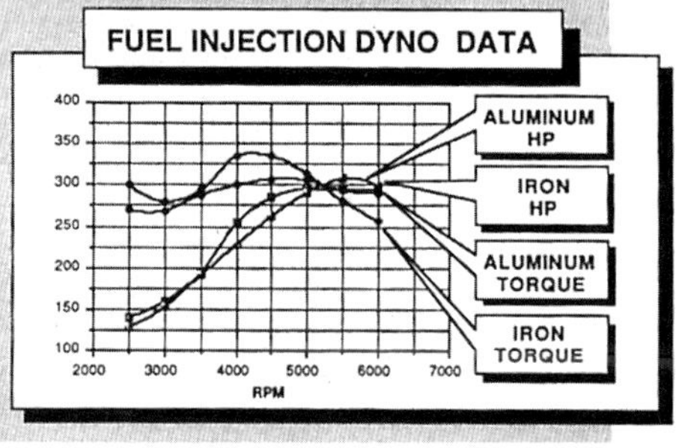

A unique spark-plug seat accepts either tapered-seat "peanut" spark plugs or the conventional large-hex, gasketed variety.

The new head has passages and hook-ups provided to accommodate stock smog pump rails, EGR, exhaust heat, and other emissions add-ons.

SOURCES

KAUFMANN PRODUCTS, INC.
Dept. HR
12400 Benedict Ave.
Downey, CA 90242
213/803-5531

TFS/TRICK FLOW SPECIALTIES
Dept. HR
6077 Back Orville Rd.
Wooster, OH 44691
216/682-8331

THE DUKE OF WINDSOR

A COMMONER SMALL-BLOCK FORD WOULD NOT DO FOR OUR PRINCELY PROJECT, SO WE COMMISSIONED JACK ROUSH TO FIT OUR STEED WITH . . .

By Gray Baskerville

Pity the poor 4-speed in our Mustang II, for it shall inherit 400 foot/pounds of torque at 4800 rpm, 400+ horsepower at 6500 rpm and all the other pressures that are associated with playing behind one, very special, Windsor Ford. As you may have guessed, that stocker 302-inch V8 had to go. It was too much a by-product of today's limited expectations. It, by gad, had to be decidedly different, befitting those performance enthusiasts who are not feint of heart. Feint of heart? Well, like ol' "Cactus" Jack Roush slyly said, "You've got to think about all those parts whizzing around in there at seven grand." What parts? Ah, it's like shades of yesteryear. Our once-docile 302 has been replaced by the hot rodder's answer, a full-boogie poke 'n' stroke number with just enough off-the-shelf parts to make the "now" 363-inch, thin-wall Hank a relatively reasonable answer to those who would like to do the same and turn their own mounts into the torque of the town.

TACKLING THE BLOCK AND RECIPROCATING ASSEMBLY

As we all know, the cylinder block is the foundation of any internal combustion engine, and to ensure that the engine will live up to expectations, it is "blueprinted," a name for making all its vertical and horizontal planes conform to a set of predetermined specifications. The 'Stang's block was no exception. Roush gave it a thorough align honing, milled the decks, bored and honed the cylinder walls, and notched the bottom of the bores. Notched the bottom of the bores? Well, that's what you need to do when you're running a stroker crank and long rods. All right! That's right. To pick up a quick 30 inches, Jack slipped a cast nodular iron 351 Cleveland crank—which has had its overall length reduced, its cam sprocket changed, the rod journals cut down .080-inch, .125-inch taken off the mains, rear seal resurfaced, oil holes enlarged, additional balancing weight added to both the counterweights, flywheel and front dampener and a full Tufftride treatment—into the 302's main webs. To the Moldex-modified cranks, eight Boss 302 Trans Am forged steel rods sporting polished and peened beams, rebuilt big and little ends and full-floating pins were attached. The rods were then fitted with a special set of .060-inch-over, flat-top BRC forged aluminum pistons encompassing a G&R-designed skirt and relocated pin bores to move the pin closer to the bottom ring land. Last, one of the specialties of the house was to take the 1/16-inch ductile chrome moly top ring, the 1/16-inch cast-iron scraper and the three-piece oil ring and calibrate their tension to ensure full oil control.

CAMMING IT UP

After a thorough search, Jack decided to rely on one of Ford's old

musclecar cams, a relatively mild hydraulic whose specs are a .470-inch lift (both intakes and exhausts), 290-degree duration and 62 degrees of overlap. Cam is run either straight up or advanced one degree, and relies on a combination of a stock Ford timing chain, Cloyes iron top sprocket and a G&R fully adjustable crank sprocket. According to the man from Livonia, this cam should come in like gangbusters at three grand and leave at around five.

OIL IN THE FAMILY

As you recall, Jack reworked the crank's oil holes, but other than a couple of other mods, he left the Windsor's lubrication alone. The first exception was the installation of one of his high-pressure, high-volume pumps and heavy-duty pump drives. The second was the reshaping and deepening of the sump in the oil pan so it would clear the crossmember. And by adding a couple of inches to the bottom of the sump, he was able to pick up an additional two quarts of oil capacity.

HEADING UP THE VALVETRAIN

The valvetrain is strictly off-the-shelf, G&R good stuff. Lifters are their anti-pump-up units, plus G&R's 5/16-inch chrome moly pushrods, G&R titanium retainers and a set of G&R valve springs. These "wounderful" items possess both inner and outer springs, plus a dampener, and provide 120 pounds of seat pressure. As for the valves, they're TRW Chevy types with a head diameter of 1.940 inches on the intakes, 1.620 inches on the exhausts. Naturally, the 351 Windsor cylinder heads have been fully ported, polished, cc'd (62.4cc),

Stock 302-inch Windsor block was align-honed, decked, bored .060-inch over, and honed with a torque plate. The bottoms of the bores were machined and notched in four places to provide the necessary clearance for the longer connecting rods and rod bolts.

PHOTOGRAPHY: JOHN O'CONNELL

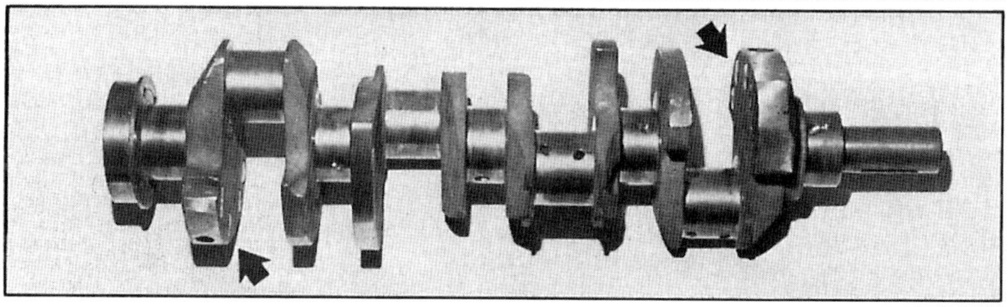

Moldex-modified 351 Cleveland crank was also given the full treatment. Length was reduced, cam sprocket changed, rod journals cut down .080-inch, 125-inch on the mains. Arrows point to Mallory.

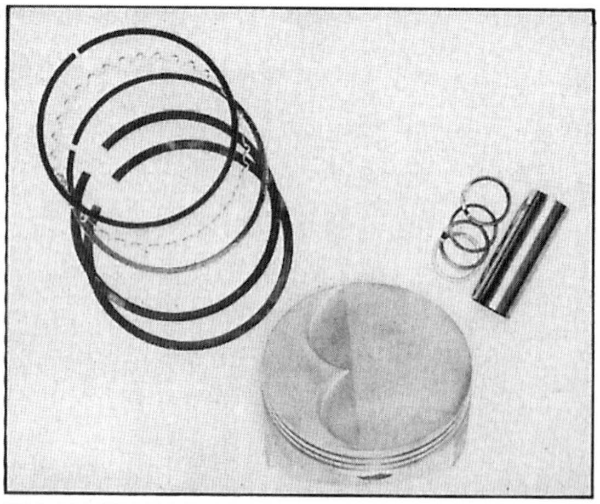

RIGHT—A set of 9.0:1 BRC forged aluminum pistons are used in conjunction with 2½-inch full-floating chrome wrist pins. End gap on the rings are .015-inch top, .012-inch scraper and oil.

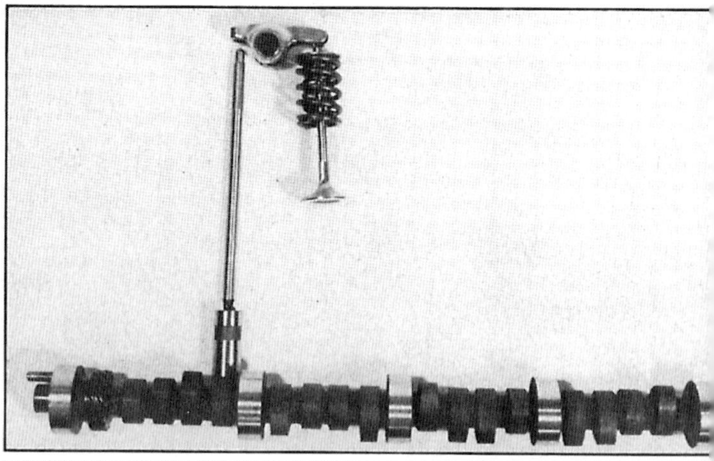

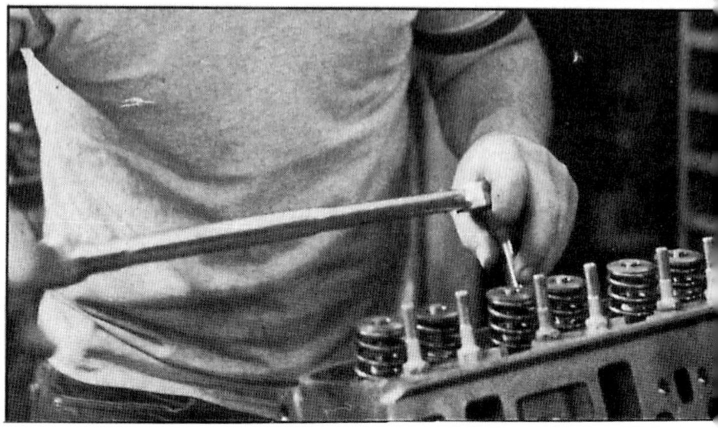

BELOW—Roush selected an old Ford muscle cam, (part No. 690Z6250C) hydraulic, and their unit pump-up lifters, 5/16-inch pushrods, aluminum rockers, titanium retainers and G&R's No. 3006 valve springs. Valves themselves are 1.940-inch intake, 1.620 exhaust. **BOTTOM—**Heads were torqued to 75 pounds max.

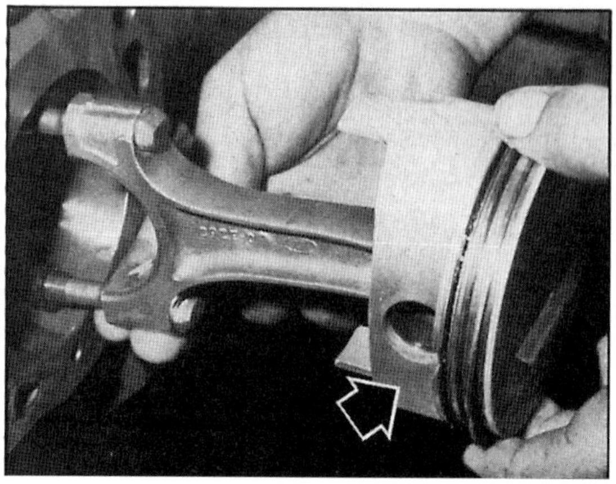

Note that the pin boss has been moved inside the bottom ring land to provide more meat around the boss. Naturally the forged steel Boss 302 Trans Am rods have been totally rebuilt. These rods (part No. DOZX6200Z) are 5.315 inches long, as opposed to the shorter (5.165-inch) stock rods.

and had the valve seats remachined to compensate for the oversized valves.

INDUCTION CENTER

As you might have guessed, ol' Cactus is still toying around with intake manifolding. He has boiled it down to either a single four-barrel, dual-plane Shelby Cobra or a fully flogged (to compensate for the increased air flow) 289 Edelbrock Torker. As would be expected, the runners were matched to the intake ports and the 750-cfm Holley double-pumper will retain its choke but will have its fuel curve reworked.

TUNE IT ON

Yes, a set of tube headers will be found inside the engine compartment, as will a high-capacity water pump and some kind of crossflow-type radiator. But because of today's rather anemic gasoline, the reworked stocker ignition will reach a total of 38 degrees at 3600 and fire through a set of Autolite BF 32s or 42s gapped at .030-inch.

As you can see, our horsepower of a different concept should make our project Mustang II more than just another pretty face. There should be enough oats there to long-leg us over to Indy in style and enough docility to get us back—that is, if the 4-speed doesn't give up the shift. **HR**

As you can see, the sump portion of the 302 Windsor has been reworked and deepened to increase the capacity of the pan from five to seven quarts.

BELOW—Stock 302-inch Windsor cylinder heads were ported, polished and cc'd (62.4). In addition, the valve seats were remachined to accept the oversized intake and exhaust valves.

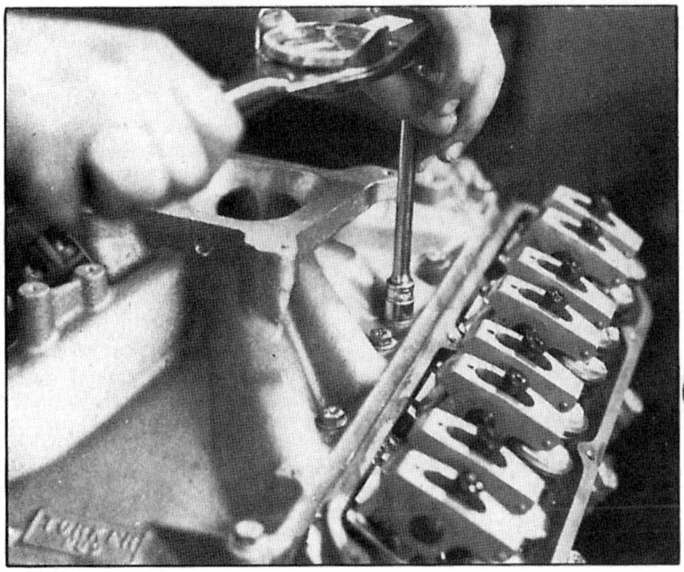

Due to the dramatic increase in the 302's displacement, the internal dimensions of this Edelbrock Torker intake manifold were considerably altered. Bolt torque specs are as follows: ⅜-inch bolts torqued to 20 pounds max, 5/16-inch bolts torqued to 10 pounds max.

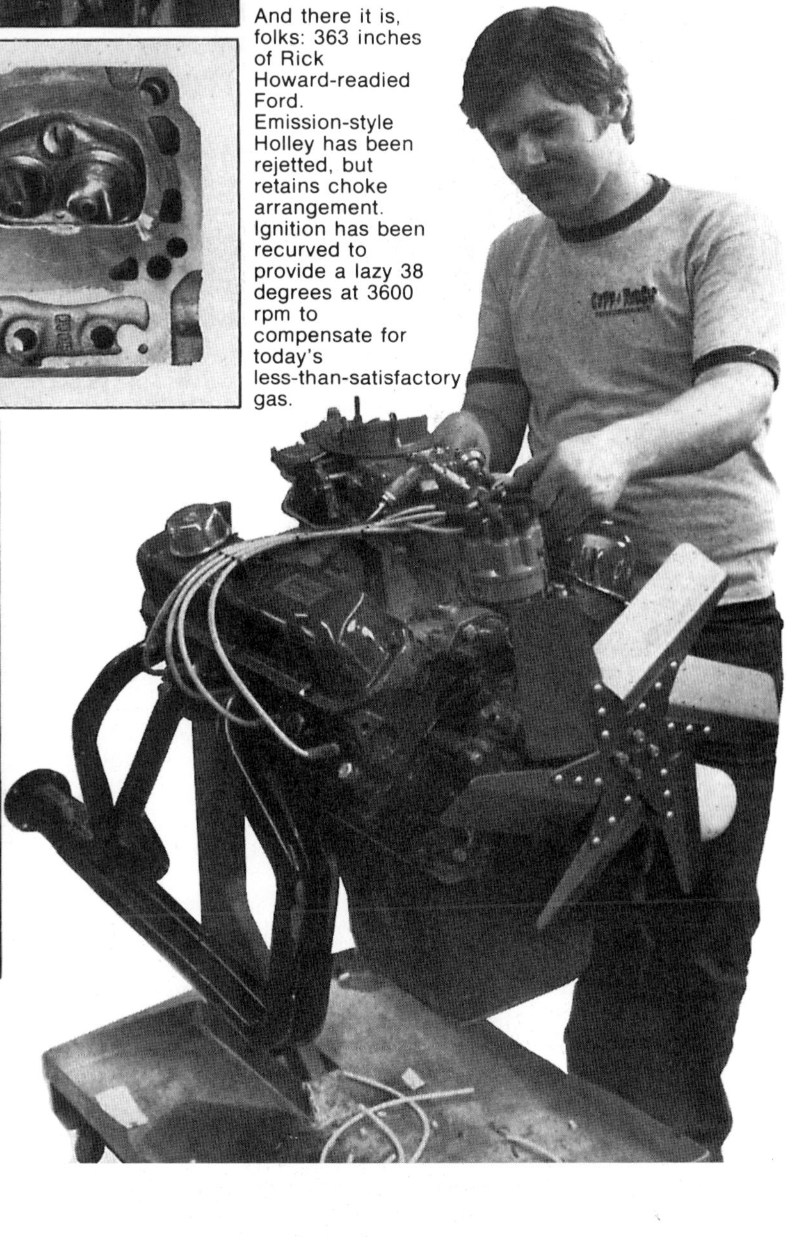

And there it is, folks: 363 inches of Rick Howard-readied Ford. Emission-style Holley has been rejetted, but retains choke arrangement. Ignition has been recurved to provide a lazy 38 degrees at 3600 rpm to compensate for today's less-than-satisfactory gas.

SORTING THROUGH THE MAZE OF FORD PARTS FOR A LOW-BUCK, HI-PO SMALL-BLOCK

By Rob Kinnan

Chevy guys have it easy. Their parts all interchange, the aftermarket is overrun with trick stuff, and every possible combination has been tried and blown-up at least a million times. Life is not so easy if your attention is focused on Ford powerplants. At times it seems that nothing interchanges, even among the same engine size and family. And looking at some of the strange designs and head/block combinations that have found their way under Ford hoods, you'd think the engineers were fond of weird experiments. Making good horsepower with stock Ford small-block parts is not impossible, but it takes some careful planning and judicious parts swapping. The Bow-Tie builders can just slap together a bunch of parts and run 11s, but that's too easy. If you like a challenge, then try it with a Ford small-block.

There are five variations of engines that can be defined by the moniker "small-block Ford." First up is the most common small-block, the 289/302 V8. Second is the 351 Windsor, which is similar to the 289/302 except that the 351 has a taller deck and a beefier bottom end. Both of these engines are known as Windsors, and they share basically the same inline-valve heads. The third small-block is the 351 Cleveland, which uses a totally different block and two different kinds of canted valve heads—two-barrel and four-barrel versions. As the name implies, the four-barrel heads were used on four-barrel–carbed engines, and the ports are way too big to make any kind of streetable torque—most street engines use the smaller two-barrel heads. The fourth and fifth small-block Fords are the Boss 302 and the 351M/400 engines. The Boss 302 combined the Windsor 302 short-block with a modified version of the massive Cleveland four-barrel heads. It was only produced in '69 and '70 and shouldn't be considered for anything other than a full-on resto. The 351/400 Midland used Cleveland heads on a tall-deck, large journal Cleveland-style block, and is also undesirable from a performance standpoint because the motor mounts and bellhousing bolt pattern are totally different than all the other small-blocks, and there are very few high-performance parts available.

If you're on a budget, build what you already have parts for. If you already have valve covers, an intake, headers, and other pieces for a Windsor, it will be much less of a hassle to build a Windsor and use those parts than scrounge for all the peripheral Cleveland parts. Because of their small displacement, the 289 and 302 will require quite a bit more effort (and rpm) to make good power, so if you're in the market and have a choice, go for the cubes and get the 351W. If you're starting completely from scratch and have no parts whatsoever, your choice should be either the 351 Windsor or Cleveland. There are pros and cons for both engines, but in most cases the Windsor is the better choice for several reasons.

Right out of the box, the 351 Cleveland with two-barrel heads will outperform a 351 Windsor, and for *mild* performance increases, it usually costs less to build a Cleveland. But, with good heads, bigger valves, and some fine-tuning, the Windsor can make excellent power. The Windsor block has thicker cylinder walls, allowing for a bigger bore, and there are several kits on the market to make a 400-plus-inch stroker out of a 351 Windsor. If your plan is to start with the basics and add on as money allows, the Windsor has more long-term potential.

Building a strong Ford on a budget uses the same strategy as most

BUDGET POWER COMBOS

The bottom end of a small-block Ford is very strong. The only thing required is throwing away the wimpy stock rod bolts and replacing them with good aftermarket bolts. The 289/Boss 302 rod on the left is a little longer than the stock 302 rod, and can easily be installed with FPP's 2991 piston for a better rod-to-stroke ratio.

These are the Windsor heads, in order from left to right: '66-vintage 289, Hi-Po 289, '70 351W, and World Products Windsor. The Hi-Po 289 heads are a waste of money—the only difference between them and the regular 289/302 heads are screw-in studs and special bosses around the valvesprings. The 351W head has a bigger chamber and slightly bigger valves than the 289 head. The World Products Windsor comes with 2.02/1.60-inch valves, better chambers, and much better intake and exhaust ports. For a 289/302, World Products also has the Windsor Jr. heads, which work very well on the smaller motor.

At far left is a TRW L2482 forged flattop, which will work with stock valves. If oversized valves are used with any performance cam, valve-to-piston clearance gets close with this piston, so FPP has a line of modified TRWs, shown here. In order from left to right (after the TRW) are FPP 2982, FPP 2991, and FPP 2986. The 2982 has bigger reliefs to clear the big valves, the 2991 has similar reliefs with more material taken off the top to allow the longer 289 rod to be used, and the 2986 started life as a TRW L2249 domed piston before FPP hacked off the dome and cut deeper reliefs.

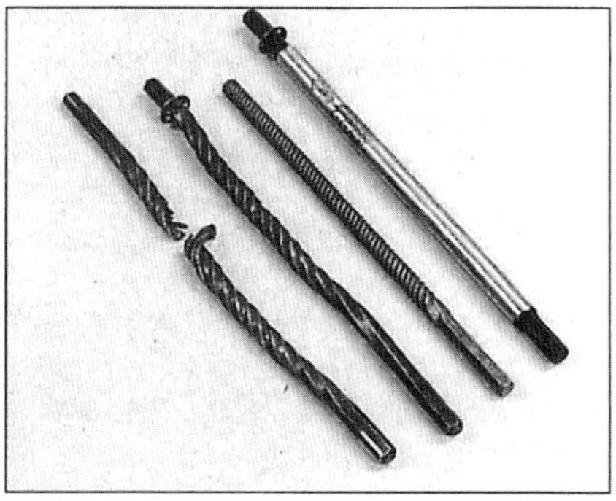

The weakest link in the Ford is the oil pump driveshaft. These shafts have twisted up like pretzels and broken. FPP has a much stronger shaft *(right)* that solves this problem.

SMALL-BLOCK FORD

buildups—the power is in the cylinder heads and induction system, not in the short-block. Money spent on a forged crank and a four-bolt–main block could be better spent on the heads for most street applications. Carefully assembled stock parts are amazingly strong and will live. Ron Miller at Ford Power Parts (FPP) in Norwalk, California, tells us there are only two weak links in the short-block, the rod bolts and the oil pump driveshaft, and both should be replaced with good aftermarket parts. One trick that's worth the effort is to install longer 5.155-inch 289 rods (casting No. C3AE) in place of stock 5.090-inch 302 rods (casting No. C80E-A) for a better rod-to-stroke ratio. Journal diameters are the same for both types of rods, and FPP has inexpensive forged pistons to make it happen.

Small-block Ford cylinder heads merit a 100-page story just by themselves, so we won't get into all of them here, just the most common. The '69-'70–vintage 351W head has decent ports and valves and a small chamber compared to the 289/302 and can be swapped back and forth between Windsor engine sizes. The bad part is that these heads are still restrictive and make it difficult to make more than one-horsepower-per-cubic-inch. The intake ports are small, the exhaust ports are *way* too small, and the valves are equally puny. For any kind of performance in any small-block (especially the 351), the valves should be swapped for bigger 1.94 intake/1.60 exhaust valves, and the exhaust ports and bowls should also be opened up as much as possible. Unfortunately, even after all this work, the Windsor head will still not flow very well, and you'll have almost as much money sunk into them as you would had you bought one of the better aftermarket heads, such as the World Products Windsor or the new TFS iron head. And out of the box, these aftermarket heads will murder even heavily ported stockers.

A Cleveland also has a problem with heads; the four-barrel intake ports are way too big, and the two-barrel heads' combustion chambers are too big for good compression. In this case, it's better to err on the conservative side and choose the open-chamber two-barrel heads, as they will make much better torque. An advantage of the four-barrel, closed-chamber heads is that they allow about ½-point higher compression ratio due to a more efficient chamber design.

A low-buck swap that works well is adapting two-barrel Cleveland heads to a 351 Windsor block. It's not a true bolt-on, but it's close. The Cleveland passes water through the block, whereas the

Windsor passes it through the intake manifold, so the water jacket opening on the face of the Cleveland head must be welded or plugged and a new hole drilled in the head's intake manifold surface. The difference in deck height between the two engines also requires a custom intake manifold, available from B&A Ford Performance, the pushrods have to be longer, the pistons need Cleveland-style valve reliefs, and Cleveland head bolts are required.

Even after all this work, it's a push as to whether this combination will make more power than a Windsor with good aftermarket heads. The aftermarket heads also use stock valve covers, headers, etc., and there are many more intake manifolds to choose from.

Camshaft timing with a Windsor is critical to performance, especially if the engine has production heads. A dual-pattern cam with 7 to 10 degrees more exhaust duration is mandatory with stock heads, and still a good idea even with aftermarket heads. The same goes for a Cleveland—add about 10 degrees of timing to the exhaust side.

Any yahoo can assemble a powerful Chevy engine, but it takes a real die-hard to coax good power out of a small-block Ford. It requires more effort, more thinking, and usually more money, but it's all worth it when that Blue Oval mill makes the Chevy guys eat a big plate of crow! **HR**

The exhaust ports of the World Windsor head *(right)* are dramatically better than the stock heads.

Mounting a 351W head to a 289/302 requires that a ⅝₂-inch water passage be drilled in the head face. Use a head gasket to line it up. Flanged head bolts or washers are also required to mount this head to the smaller motor.

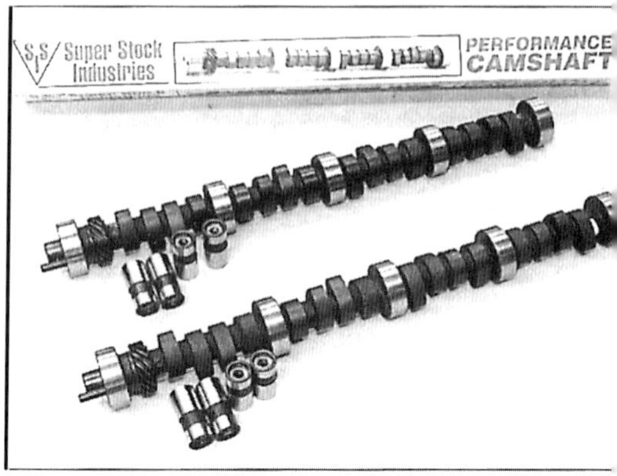

A Windsor-headed small-block needs more cam timing on the exhaust side to make up for the restrictive exhaust ports. A Cleveland follows the same basic guidelines. Performance Automotive Wholesale has a line of Super Stock Industries cam-and-lifter combos that are priced right for the budget-minded. Regardless of how much lift your cam has, *always* check valve-to-piston clearance, especially if you've installed bigger valves.

The two-barrel Cleveland intake port on the left flows well enough for most street engines, and is always a better choice than the four-barrel head. You can almost stick your whole fist in the big four-barrel port on the right! There's also an Australian Cleveland head that features the two-barrel intake ports but the four-barrel's closed combustion chambers. They're harder to find and more expensive, but in some cases are worth it.

SMALL-BLOCK FORD

	351W	351C
CARBURETOR		
size	750 cfm	750 cfm
type	4 bbl with vacuum sec.	4 bbl with vaccum or mechanical sec.
INTAKE MANIFOLD		
type	dual-plane	high-rise dual-plane or single-plane
CAMSHAFT		
type	hydraulic or solid	hydraulic or solid
intake duration @.050-inch lift	215-235 degrees	220-235 degrees
exhaust duration @ .050-inch lift	222-245 degrees	225-240 degrees
gross intake lift	.490-.530 inch	.500-.550 inch
gross exhaust lift	.500-.550 inch	.510-.560 inch
lobe separation	110-112 degrees	110-112 degrees
HEADS		
type	ported 351W or aftermarket	2-bbl Cleveland
intake valve size	1.94-2.02 inches	2.19 inches
exhaust valve size	1.6 inches	1.71 inches
COMPRESSION		
static	8.5-9.5:1	9.0-9.5:1
EXHAUST		
headers:		
tube diameter	1⅝ inches	1¾ inches
collector diameter	3 inches	3 inches
exhaust system and muffler inlet/outlet size	2¼-2½ inches	2½ inches

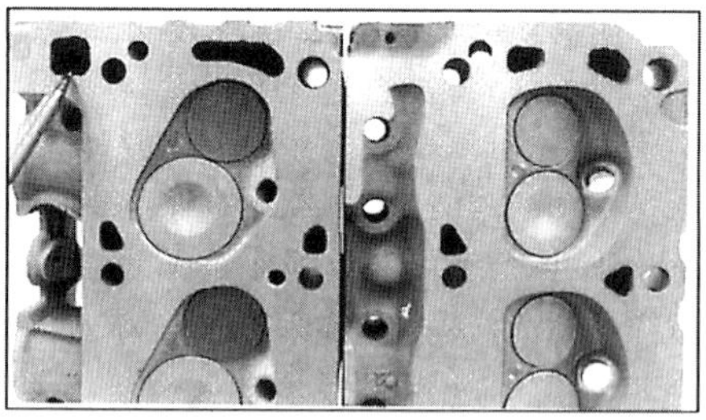

Clevelands pass water to the head through the block and Windsors get it from the intake manifold. When mounting a Cleveland head *(left)* to a 351 Windsor, the hole pointed to here must be welded closed or machined to accept a press-in plug, then a new hole has to be drilled on the intake mounting surface. This is a good low-buck swap but it requires a custom intake manifold. B&A Ford Performance makes both dual-plane and single-plane intakes for this swap on either a 289/302 or a 351W. The B&A manifolds can also be ported to fit four-barrel heads, and they cost around $250.

SOURCES

B&A Ford Performance
Dept. HR03
P.O. Box 6553
Ft. Smith, AR 72906
918/427-7757

Ford Power Parts
Dept. HR03
14504 S. Carmenita, #C
Norwalk, CA 90650
310/921-5300

Performance Automotive Wholesale
Dept. HR03
21050 Lassen St.
Chatsworth, CA 91311
818/998-6000

Trick Flow Specialties, Inc.
Dept. HR03
P.O. Box 593
259 E. Main St.
Smithville, OH 44677
216/669-2838

World Products, Inc.
Dept. HR03
353 Oliver St.
Troy, MI 48084
313/244-9822

BLUE OVAL TECH

FUN WITH FORDS

AVOIDING THE 10 MOST-COMMON FOMOCO FOIBLES

By Jeff Smith

It takes dedication to be a hot rodder. You have to be part engineer, part physicist and part mechanical genius. If you happen to be a Ford fan, you must learn to avoid the common pitfalls that are a no-charge part of the Blue Oval world. It must be that Ford engineers take deranged pleasure from making parts that *almost* interchange. While hot rodding will always be a game of mixing and matching parts, it seems especially tough for Ford fans.

Oftentimes, it's helpful for a hot rodder to know what *not* to do, or at least where the problems lie so he can either fix or avoid them altogether. To help you out, we've talked to a few knowledgeable Ford sources such as John Vermeersch at Total Performance (who also runs the Motorsports Technical Hotline), Chris Kaufmann at Kaufmann Products and Mark Sanchez at Advanced Engineering West to come up with 10 of the most common Ford mistakes and ways to avoid them. So commit these gems to memory and make hot rodding those Fords a far more fun experience.

GEAR UP

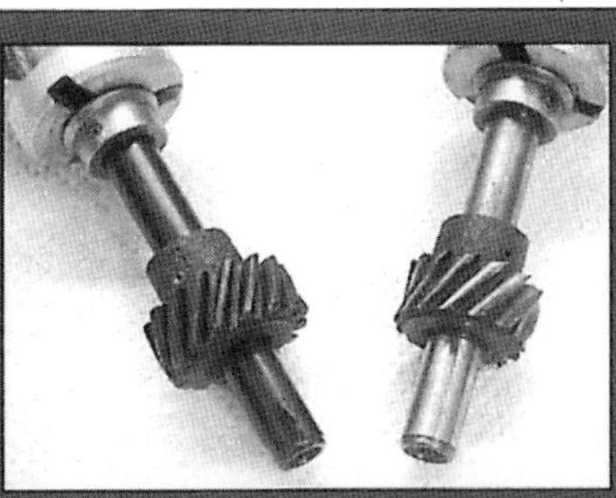
The distributor gear on the right illustrates what happens when you don't use the proper distributor gear with a steel roller cam. Imagine all those iron filings going straight into the oil!

The late-model, roller-cammed, fuel-injected 5.0L engine has become the powerplant of choice for a number of Ford hot rodders. The popular plan is to buy a used fuel-injected engine out of a wrecked Mustang or T-Bird and fit it with a carburetor and a different distributor. What most rodders don't realize is that the steel roller cam requires a special distributor gear. Since all 289/302 Windsor Ford distributors interchange, it's easy to drop in a distributor and fire up the engine. Unfortunately, the steel roller cam will quickly chew up a nonroller-cam distributor gear and will probably destroy the cam gear as well. The quick fix is to add a bronze or specific steel-compatible distributor gear to the distributor. Crane Cams offers a high-quality silicone copper alloy gear that is superior to the inexpensive bronze gears. There are two different shaft diameters—.467 inch (part No. 52989-1) and .500 inch (part No. 52990-1)—for the 289/302 Windsor distributors, and remember that even the silicone alloy gear will wear at a faster rate than normal.

The Ford SVO catalog lists three different gears that can be used with a .467-inch distributor shaft. Its newest is a steel gear compatible with the factory steel roller cam and aftermarket cams (part No. M-12390-B), while a larger .531-inch-shaft-diameter steel gear for 302 EFI roller-cam motors and all 351W engines is also offered (part No. M-12390-F). An SVO bronze gear (part No. M-12390-E) is also available.

BALANCING ACT

Another problem that crops up when swapping later-model small-block Fords into earlier-model cars is flywheel or flexplate balance. Many Ford enthusiasts don't know that Ford changed the external balance weights of the Windsor small-block in 1980 with the introduction of the short-lived 255ci V8. The external balance changed from 28.2 ounces to 50 ounces with the '80 engines. Other than this change, the flywheel/flexplates are the same, which means that they will interchange. While this change may not sound significant, if you bolt a pre-'80 flywheel or flexplate to an '80-or-later small-block (or vice versa), the engine will vibrate horribly from the moment you fire it up, causing severe engine damage if driven for even a short period of time. In some instances, the later-model flywheel or flexplate may not fit inside the earlier bellhousings, since Ford offered both 157-tooth and 164-tooth flexplates.

This flywheel and harmonic dampener are for the heavier 50-ounce external-balance, late-model 5.0L engine. Notice the size of the imbalance weight located on the flywheel and harmonic dampener. The lighter 28.2-ounce flywheel or dampener will have roughly half the size imbalance weights, making it easy to spot.

ALL'S IN ORDER

Many neophyte Ford freaks don't realize that the firing order of the early 351W engine is different from the 221-, 260-, 289-, 302- and late 351W engines. The early 351W's firing order is 1-3-7-2-6-5-4-8, while the smaller Windsor engine's firing order is 1-5-4-2-6-3-7-8. What makes this interesting is that 289/302 cams are completely interchangeable with 351W cams, providing you match the firing order to the camshaft. If a 289/302 cam is used in a 351W engine, the firing order must be changed to the 302 sequence by moving the wires on the distributor cap. The same procedure is used if a 351W cam is used in a 289/302W engine where the firing order in the distributor cap must match the camshaft.

Unfortunately, this trick won't work with sequential EFI engines like the 5.0 if a 351W cam is installed, so be forewarned. According to Kaufmann, switching the plug wires will make the engine run, but power will be down dramatically. The fix is a new ECM to establish the correct injector firing order.

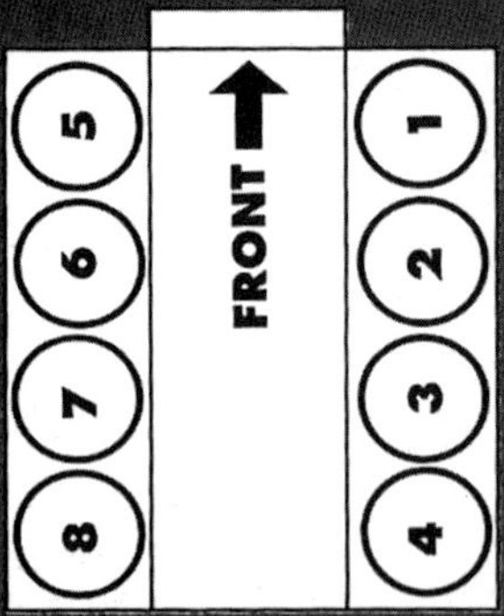

Not only is the firing order different on the 302- and 351W engines, but Ford also numbers its cylinders differently than GM engines.

REVERSE COOLING

Engine swapping late-model, serpentine-belt-accessory-drive-equipped engines into earlier cars can also lead to problems if the late-model engine is retrofitted with a standard V-belt pulley drive system. It's easy to overlook the fact that serpentine-belt drives rotate the water pump in the opposite direction from V-belt drive systems. If you equip a late-model 5.0L engine with V-belts without changing the water pump, water pump efficiency will drop radically, and the engine will overheat. The same is true if an early-model engine is equipped with a serpentine-belt drive. The simple fix is to equip the engine with the appropriate water pump for the accessory-belt drive you are using, regardless of engine year.

There's no easy visual way to tell a serpentine-belt-drive water pump from a V-belt water pump unless you remove the rear cover of the pump and look at the the pump impeller vanes. So make sure your pump is correct for the belt-drive-system application.

TIMING MIX 'N' MATCH

Hot rodders revel in the mix 'n' match form of building a car. Often, parts come from unknown or long-forgotten sources. One source of headaches is Ford's switch from three-bolt harmonic dampeners to four-bolt pulley-mount dampeners around 1968. While that can cause problems in itself, the real challenge comes when the timing marks don't line up. In addition to the pulley-bolt-pattern change, Ford also changed the location of the TDC mark from the left (driver's side) to the right side of the engine. If your timing mark is 90 degrees off, you can change pulleys, change the location of the timing mark indicator or use an aftermarket timing tape to indicate TDC referenced 90 degrees from the original mark.

This is a late-model, four-bolt harmonic dampener for a 5.0L engine. Note the position of the TDC mark (*1*) compared to the crank keyway (*2*). Earlier three-bolt dampeners will have a different location for the TDC mark. Keep in mind that earlier three-bolt dampeners will have a lighter external balance, too.

T-5 TRANS TRICKERY

The Borg-Warner T-5 trans is the hot ticket for budget Ford street heroes, but the proliferation of those transmissions can lead to problems. Before you step up to buy a used T-5, be aware that not all T-5s are the same. Later-model Mustang T-5s, beginning with '90 Mustangs, are upgraded transmissions that offer much stronger nickel steel gears and superior synchros and are highly prized by knowledgeable Ford fans. They can be identified by a B-W tag listing either a 1352-208 or 1352-199 serial number. If you are junkyard or swap-meet hunting, keep in mind that the four-cylinder T-5 transmission looks the same but actually has a longer input shaft and slightly smaller-diameter pilot shaft. If you bolt the four-cylinder T-5 into a V8 application, the clutch will chatter on engagement because the pilot shaft is loose inside the pilot bushing. Worse yet, the longer input shaft presses forward into the crankshaft, placing pressure on the thrust surface of the main bearing, which will quickly destroy the bearing and cause major engine damage. If that isn't enough, four-cylinder transmissions have deeper 4:1 First-gear ratios that are substantially weaker than the V8 boxes. Borg-Warner engineer Steve Whitaker advises that you steer clear of pre-'85 V8 T-5s as well since they are a little weaker than the post-'85 models. Furthermore, the new '94½ Mustang T-5 is different from either of the previous T-5s and will not interchange. Hey, nobody said it was easy! The bottom line is that there's no reason to buy a four-cylinder T-5, even if it's dirt cheap.

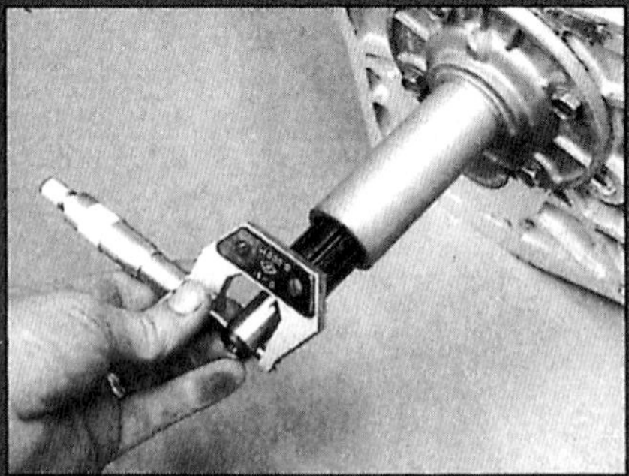

TRANSMISSION	Pilot Shaft Diameter (inches)	Total Input Shaft Length (measured from trans face, in inches)
4-Cyl.	.590	7.41
V8	.668	7.15

LEAKPROOF

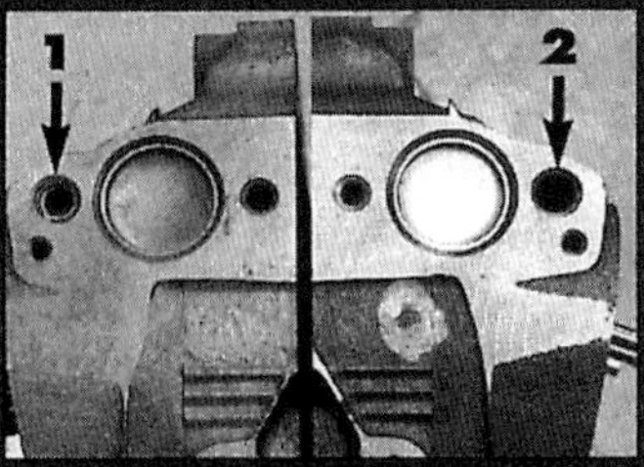

The head on the left has the Ford-supplied insert (*1*) that also blocks off the exhaust port. The head on the right is open (*2*). You need to look carefully to ensure that the plugs are in place, or you'll have a noisy exhaust leak.

All Ford small-block heads since the mid-'70s are configured to accept smog-pump input into a hole in the back of the right cylinder head that is plumbed to the exhaust ports. The hole is drilled and tapped in both heads for a ⅝-inch national coarse (N.C.) thread. Ford inserts an adapter plug that is internally drilled and tapped for a ⅜-inch N.C. thread for the opposite end of the head in the front to mount the accessory-drive bracket. When an engine is rebuilt and bolted into an emissions-exempt vehicle, the rear hole may be left unplugged. Left open, the holes will generate a tremendous exhaust leak that may not be immediately apparent and could be blamed on leaky header gaskets. If the smog pump is not used, the O.E.M. Ford ⅝-inch blind hole insert (Ford part No. 351418-S) is a simple and inexpensive fix to this problem.

HEAT TRICKS

Late-model Mustangs often suffer from overheating problems that can be caused by a number of factors. Properly filling the cooling system is difficult in fuel-injected Mustangs because a portion of the upper radiator hose is higher than the radiator fill cap. Many Showroom Stock road racers drill a pair of ⅛-inch holes in the thermostat to help alleviate the air pocket that can form in the hose.

Drilling two ⅛-inch holes in the thermostat can help eliminate air pockets in the cooling system, which could contribute to overheating problems.

Other helpful hints involve not changing the stock thermostat to a 160-degree model on EFI Mustangs. That can often fool the ECM into richening the mixture, which only makes the engine run worse, increasing emissions and killing gas mileage.

IT'S ALL IN THE TIMING

Sometimes it's the simple things that make a huge difference. All EFI Ford engines require that the shunt wire between the distributor and the computer be disconnected before the initial spark timing is set. The shunt is located right beside the distributor on all EFI V8 Mustang and T-Bird engines. If it is not disconnected, a timing light will indicate the total timing between initial and the amount added by the computer. Setting the initial timing at a stock 10 degrees without disconnecting the shunt actually retards engine timing by as much as 12 to 18 degrees, killing engine power. If you want to try to improve power, try disconnecting the shunt and bumping the initial timing to 12 or 14 degrees. Load the engine by accelerating in Third gear and back out quickly if the engine detonates. Then pull the timing back until detonation ceases.

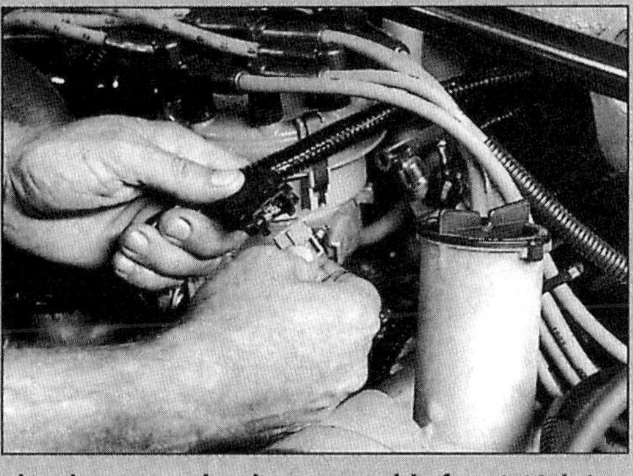

The shunt must be disconnected before initial timing can be accurately measured on all EFI small-block Fords.

ALL STEAMED UP

Using a quality head gasket, such as Fel-Pro part No. 1011, line up the steam holes between the gasket, head and block. If you find the indicated passages (*arrow*) not drilled in either the head or the block, carefully drilling them could eliminate an overheating or detonation problem.

When mixing and matching Ford small-block cylinder heads and blocks, there is a great possibility of mismatching steam-pocket relief holes between the heads and the block. If the steam pockets are not vented properly, they can cause internal water-jacket hot spots that can lead to detonation and engine damage, especially in high-output, high-compression situations. The fix for this problem is to line up a head gasket such as a Fel-Pro part No. 1011 on first the head and then the block to check for small holes located on the intake side of the cylinder deck surface. The holes should line up with holes in the cylinder heads. If there are holes in the block but not in the head, you should very carefully drill through the deck surface of the head or block to ensure that the steam-pocket points are open. Carbide drill bits work best, especially in high-nickel-content blocks. This mismatch can even occur on original factory engines and could account for engines with a sensitivity toward overheating or detonation problems. **HR**

SOURCES

Advanced Engineering West
Dept. HR12
12412 Bennedict Ave.
Downey, CA 90242
310/803-3677

Kaufmann Products
Dept. HR12
12420 Benedict Ave.
Downey, CA 90242
310/803-5531

Total Performance
Dept. HR12
44020 N. Groesbeck Hwy.
Mount Clemens, MI 48043
313/337-1357 (SVO Tech Line)

WIN-SOAR

JOE SHERMAN TURNS A WINDSOR INTO A WINNER

By Gray Baskerville

Pity us poor Ford lovers. Since 1955, when Chevrolet sprung that heartburn V8 on us, we've battled the big Bow-Tie. Ford turncoat Zora Arkus Duntov made it even worse. He suggested to Chevy that the factory offer rodders low-buck, high-performance parts so they could stick together a hot Chebby on the cheap. Ford failed to follow suit and got "blued" off. But things are looking up. One of Joe Sherman Racing's customers commissioned the Sultan of small-blocks to assemble a replacement for the ailing small-block in his 5.0 Mustang. Joe's goal was to "Shermanize" a Windsor into a nice 480- to 490-horse motor. After the smoke cleared, the 408ci stomper's final numbers even astounded Sherman—543 hp and 527 lbs-ft of twist from a 10.1:1 streeter. "Haul-lelujah," fellow Fordnatics, we've been "Dearborn" again.

What makes these power figures impressive is that Sherman can build a street-flexible, high-output, 408ci Ford for virtually the same kind of money that he would charge for a comparable 406ci Chevy. So instead of taking up any more space, let's take a look at Sherman's method of turning a Windsor into a winner.

Photography: Gray Baskerville

The 4.030x4.00-inch bored and stroked 408ci Ford is based on a '75 351 Windsor. Joe Sherman's original goal was to make 490 horsepower. However, this mild child cranked out 523 hp at 6500 rpm and 504 lbs-ft of torque at 5200 through the mufflers. It pulled 543 hp at 6250 rpm uncorked and 525 hp at 4750 rpm on Sherman's dyno.

Sherman couldn't find a bare block, so he bought a '75 vintage long-block—pistons, rods, cam, crank, and bearings—from a local recycler for $150. One thing to consider is that the front timing chain covers tend to corrode and are costly to replace. The cost-conscious Sherman bought a new one for $160.

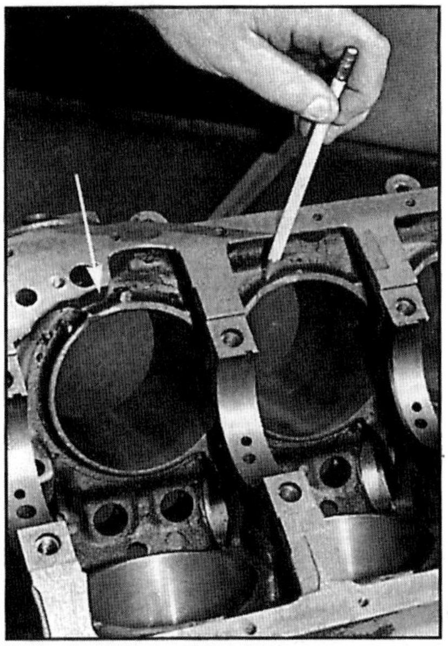

Other than the usual machine work—align honing, decking, and boring—a portion of the oil pump pad and the bottom of the bore should be machined to ensure that both the crank's counter weights and connecting rods will clear.

Another recycled item was this 351 Cleveland nodular iron crank. These increasingly popular cranks provide a 4-inch stroke. The one drawback is that its long front snout bends easily, and Sherman's crank grinder had to find a straight shaft. Cranks cost between $150 to $300, and Sherman spent an additional $350 preparing the stroker. The 3.00-inch Cleveland mains were cut down to the Windsor's dimension of 2.75 inch. The cranks can also be offset ground to increase its stroke up to 4.20, but that wasn't done because Sherman wanted to keep the compression ratio around 10:1. The outside diameter of the counterweights should also be machined so they clear the inside of the block. Castillo Crank Grinding Service, Sherman's grinder, offers this conversion for $350, excluding offset grinding.

Sherman selected an Isky RR-608 roller cam with a Chevy grind (260/270 degrees duration at .050 inch on 112 centerlines with .642-inch lift on the intake and .630-inch lift exhaust) on a Ford blank that Isky had never produced before. According to Sherman, "I picked a nonaggressive roller that is one or two notches above a street grind, which comes out a little bit bigger than that of a comparable Chevy blank. The lower timing chain sprocket features nine key-way positions, which allow the cam to be retarded or advanced. Sherman runs his between 2- and 3-degrees advance for improved flexibility.

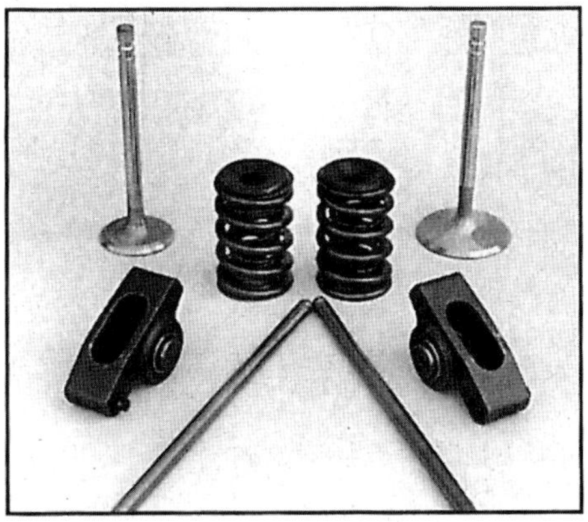

The valvetrain is common street fare, featuring 2.02-inch intake valves, 1.60-inch exhausts, and double-wound springs. The pushrod length must be verified, but stock length could potentially work.

PISTON PROTOCOL

Piston selection is based on deck height and stroke length, Sherman decided the static compression should be 10:1. He juggled the rod length with piston pin height to arrive at that number. After a search through his parts manual, Sherman came up with a mix that combined 6.125-inch 360 Chrysler rods with off-the-shelf flattop Chevy pistons. In order to use this combo, the rod journals are ground from the stock Ford size of 2.311 inch to fit the Chrysler's 2.125-inch big end. The rods' small ends were rebushed to accept a stock Chevy .927-inch pin. Now Sherman could use a common Manley 4.030-inch replacement aluminum piston (part No. 49443) that would

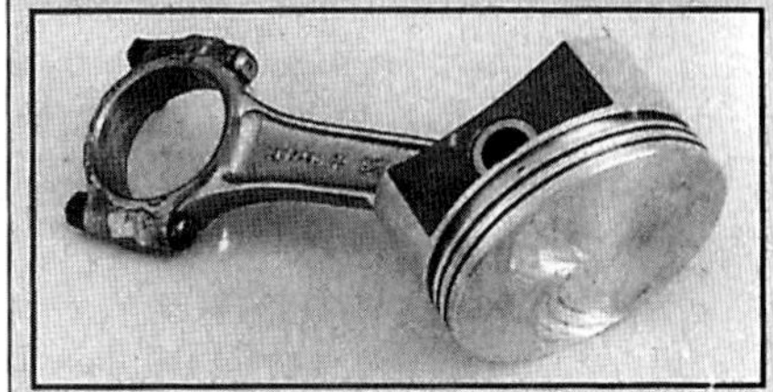

end up .060-inch down from the top of the deck. Due to valve lift, combustion chamber size, and piston height, the top of the piston may require some additional valve pocket relief work. However, there should be no valve clearance problems in low compression/mild cam applications. Meanwhile, these pistons saved the customer from having to buy custom pistons or pay for extensive crank work. And with offset grinding on the crank, the customer could have ended up with over 12.0:1 using these same rods and flattop pistons.

Sherman chose the new cast-iron Dart Windsor heads for this application. The combustion chambers proved to be a tight 60 cc's, so Sherman had to remove material to open them up to 64 cc's. Intake port flow increased from a stock 226 cfm at .600-inch valve lift to 259 cfm, while the exhaust improved from its stock 135 to 182 cfm mostly from minor bowl and short-turn radius work. Keep in mind that these heads were originally intended for 5.0L Ford applications!

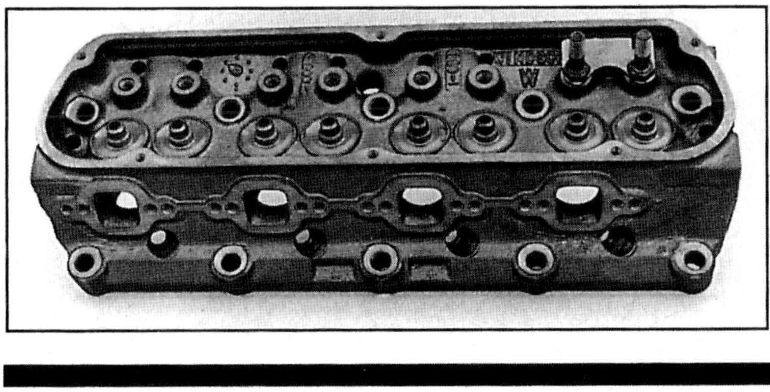

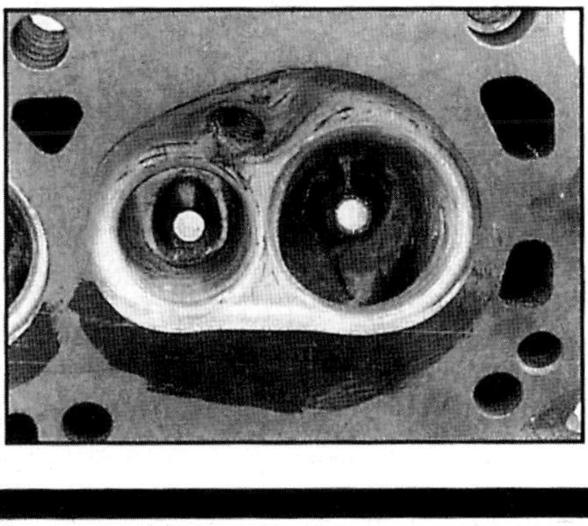

STEALTHY 'STANG

Joe Sherman's *modus operandi* is building low-profile, high-horsepower small-block Chevys. He also likes a challenge. Bob Harrison, a close friend of Joe's brother Bill, needed a Windsor to replace the fast-fading 5.0L V8 that powered his '83 Mustang. The end result was a Sherman-built, 408ci Windsor—assembled from over-the-counter parts—that out-horsed Harrison's tired 302 by a good 400 ponies. Once the new engine was finished and the dyno runs were completed, Bill Sherman and Harrison spent the next two months making the horse trade.

Aside from swapping powerplants, the trade necessitated two major changes: exchanging the original Ford trans for a GM Turbo 350 and reworking a set of prefabricated headers so that they would clear the new trans mount. The finished project has forever changed Harrison's concept of horse pressure. He now knows the meaning of the term "traction-limited" every time the secondaries open in what was once just an old, 140,000-mile commuter car.

Joe Sherman is convinced that once Harrison swaps in an 4.10-geared 8.8 rearend and a set of slicks, the Mustang will be capable of low 11's at over 125 mph at the dragstrip!

The heater and air conditioning units were lost in the swap. The Mustang, according to Sherman, "is now a weekend play thing."

Even though it's still equipped with 2.73 gears and a 3000-stall converter, there's enough Joe Sherman-built and tuned FoMoCo under the hood to break the tires loose in the first two gears.

The owner was concerned about the driveability of the engine, so the first dyno test was done with a Sherman-modified 780-cfm vacuum-operated secondary Holley carb. The first pull through the mufflers resulted in a peak figure of 490 horses at 6000 rpm and 470 lbs-ft of torque at 5000 rpm. With a quick tune-up, those numbers improved to 513 hp at 6000 rpm and 490 lbs-ft at 5000 rpm. Sherman then switched to an 800-cfm double pumper and power increased to 523 hp at 6500 rpm and torque to 504 lbs-ft at 5200 rpm. The mufflers were then removed and the 408 engine instantly picked up 20 horses and 22 lbs-ft of torque.

The Edelbrock Victor Jr. 351W intake was cleaned up by grinding and radiusing the runners in the plenum chamber and matching the intake ports to the cylinder heads.

One trick is that Sherman uses Chevy rockers—1.6 on the intake, 1.5 on the exhaust, and the bigger $\frac{7}{16}$-inch studs.

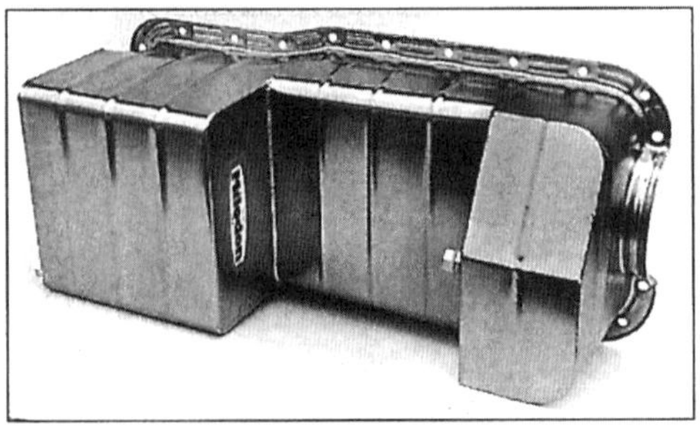

A radically notched Muldoon rear sump pan was used in conjunction with a stock Ford oil pump fitted with a custom pickup. **HR**

SOURCE

Joe Sherman Racing
714/434-9110

5.0 TO GO

DYNO TESTING FORD'S REMANUFACTURED HIGH-OUTPUT 5.0L ENGINE

By Jeff Smith

Assembled engine packages have become popular items among hot rodders in the last few years. While most sources for these engines are in the aftermarket, new-car manufacturers are also interested in these engine packages, as evidenced by Chevrolet's 350 H.O. engine package. While Ford continues to develop its 351 Windsor H.O. response, Ford Authorized Remanufacturers (FAR) are already in the hunt with a late-model Remanufactured High Output (RHO) 5.0-liter engine that's available through a number of sources, including select Ford and Lincoln-Mercury dealerships and Ford Motorsport dealerships.

As the RHO 5.0-liter name suggests, this engine is not new, but a remanufactured small-block that features a number of power-producing modifications. The RHO 5.0 comes as a complete, assembled long-block with heads, all the valvetrain components (including the timing chain and gears), and an oil pump. It does *not* include the oil pan, timing chain cover, water pump, intake manifold, exhaust, or distributor.

The RHO was developed with a couple of interesting departures from the normal rebuilt-engine approach, such as the cam FAR chose to use. With assistance from Ed Lyons and John Vermeersch at Ford's Special Vehicles Operations (SVO), the RHO engine is fitted with the Ford Motorsport M-6250-B303 roller camshaft. With specs of 284-degrees advertised duration, 224 degrees at .050-inch tappet lift, a .480-inch valve lift with the stock 1.6:1 rocker ratio, and a 112-degree lobe-separation angle, this cam is dramatically stronger than the stock H.O. 5.0-liter cam. As you might assume, it's a little lumpy when it comes to idle

Shown is FAR's RHO engine straight from the factory. It's a complete long-block minus details such as an oil pan, timing chain cover, water pump, intake, flywheel and balancer, and other necessities. These pieces are left off to reduce the initial purchase price, since most enthusiasts will be using this engine to replace an existing Windsor small-block.

Details worth noting include the fact that this is a late-model 5.0-liter engine that requires the use of a late-model balancer and flywheel or flexplate. A McLeod flywheel (part No. 463106) was used for the dyno tests, along with a stock late-model balancer and a 50-percent underdrive V-belt pulley (part No. 5455) from B&B.

Here is the RHO engine in the "stock" configuration test. It was assembled complete with a set of late-model Ford stainless-steel "headers" from a Mustang and a true 2¼-inch dual-exhaust system feeding into a pair of turbo-style mufflers.

quality. Our sources claim that it is not recommended for engines equipped with the early Ford speed-density electronic-fuel-injection system, unless you use the Ford Motorsport Mass Air Flow conversion kit. The idle quality on our carbureted dyno engine revealed 13.5 inches of vacuum at 850 rpm.

The other major modification made to this engine is significant port-work to the exhaust side of the heads. As any Windsor Ford fanatic knows, the exhaust side of most small-block Ford heads is weak in comparison with the intake ports and constitutes an area for improvement. The cast-in EGR bump in the port roof of these otherwise-stock heads is completely removed by the rebuilders, which appears to significantly improve the exhaust-port flow and to complement the potential of the B-303 camshaft.

TEST 1

A stock-configuration RHO 5.0L engine with a Holley 650 double-pumper, stock '85 Mustang factory intake, stock factory stainless headers, 2.25-inch dual-exhaust system with turbo mufflers, Ford electronic distributor, MSD-6A amplifier, MSD wires, and 36 degrees of timing.
* Asterisk indicates peak numbers.

RPM	TORQUE (LBS.-FT.)	HORSEPOWER	VOLUMETRIC EFFICIENCY (PERCENTAGE)
2000	262	100	86
2250	263	113	89
2500	264	126	91
2750	275	144	90
3000	284	162	91
3250	293	181	93*
3500	296*	197	93
3750	295	211	92
4000	291	221	92
4250	284	230	91
4500	276	237	89
4750	267	242	88
5000	254	242	86
5250	244	244*	85
5500	231	242	82
5750	215	236	78
6000	200	228	76

TEST 2

All equipment is the same as in Chart 1 with the addition of Hedma 1⅝-inch primary-tube headers.
Note: Torque gains are based on the previous test.

RPM	TORQUE (LBS.-FT.)	HORSEPOWER	VOLUMETRIC EFFICIENCY (PERCENTAGE)	TORQUE GAIN(+)
2000	270	103	89	+ 8
2250	271	116	91	+ 8
2500	279	133	93	+ 15
2750	288	151	94	+ 13
3000	300	171	94	+ 16*
3250	307*	190	95*	+ 14
3500	307*	205	95	+ 11
3750	302	216	94	+ 7
4000	300	229	93	+ 9
4250	291	236	92	+ 7
4500	282	241	91	+ 6
4750	275	249	88	+ 8
5000	263	250	87	+ 9
5250	253	253*	86	+ 9
5500	238	250	83	+ 7
5750	221	242	80	+ 6
6000	205	235	77	+ 5

The remainder of the engine is a close-tolerance rebuild of a select late-model, roller-cam H.O. 5.0-liter small-block, featuring .030-inch overbore (making this engine 306 cubic inches), low-tension oil rings, forged four-eye-brow pistons, a reground H.O. crankshaft, rebuilt H.O. rods with ARP rod bolts, and a Ford Motorsport oil pump. The heads are also completely rebuilt with new 1.78-/1.48-inch valves, guides, seals, steel retainers (without the rotators), and a 3-angle valve job. The heads on our engine were '78 70cc castings that produced an 8.5:1 compression ratio. Later-model fuel-injection truck heads are the preferred castings, as they offer smaller chambers that increase the static compression ratio.

On the intake side of this first test, we used a stock '85 Ford aluminum intake, along with a 650 Holley double-pumper. The 650 Holley was used only because we didn't have a 600-cfm-vacuum secondary carburetor at the time of the test! On the street, a 600-vacuum secondary carb would be perfect for this engine. The distributor is an '85 Ford electronic unit with a steel drive gear (part No. M-12390-F) available through SVO Motorsport. This gear is a must for any distributor, since an iron gear will wear out in a matter of hours when combined with a steel-billet roller camshaft. We also used MSD wires and an MSD-6A amplifier in all of the tests.

The RHO engine looked impressive enough on paper to warrant an in-depth look, so we enlisted the help of our friends at McFarland & Associates, who put one of these engines on the dyno to see what kind of power it could produce. Chris Kaufmann of Kaufmann Products also lent his technical assistance in the effort. In addition, we had a chance to test two intake manifolds, including Edelbrock's brand-new Performer RPM intake for the 5.0 engine and the brand-new World Products/Dart II iron cylinder head.

To begin the test, we outfitted the RHO 5.0 engine with a stock '85 Ford dual-plane intake manifold, a 650-cfm Holley double-pumper carburetor, an '85 electronic ignition distributor from Ford Motorsport with an MSD-6A amplifier and MSD wires, stock stainless-steel factory headers, and a 2½-inch-diameter dual-exhaust system with a pair of turbo-style mufflers. Once the engine was sufficiently broken in, dyno operator Kevin McClelland experimented with the total timing and jetting on the 650 Holley carburetor until power was optimized. In this "stock" configuration, the RHO engine generated 296 lbs.-ft. of torque at 3500 rpm and a respectable 244 horsepower at 5250 rpm.

Even with the enhanced exhaust ports, Kevin felt that the factory-style exhaust headers left much to be desired, so the first change was installing a set of Hedman 1⅝-inch headers, originally designed for an early small-block Fairlane. Retaining the same exhaust system and mufflers, the RHO engine responded to the change with an across-the-board increase in torque, especially below 4000 rpm where the increase was never less than seven lbs.-ft., with a whopping 16-lbs.-ft. increase at 3000 rpm! The increase in horsepower was not as significant, with a 9-horsepower increase at 5250 rpm, indicating that the factory intake was too restrictive to allow the engine to take advantage of the improved exhaust system.

Naturally, the next step was to try Edelbrock's newest high-rise Performer RPM intake. Once the intake was installed, Kevin again went through the optimization drill before coming up with a power comparison. The major increases

TEST 3

The same as the previous test with the addition of an Edelbrock RPM dual-plane high-rise intake manifold.
Note: Torque gains or losses are based on previous test.

RPM	TORQUE (LBS.-FT.)	HORSEPOWER	VOLUMETRIC EFFICIENCY (PERCENTAGE)	TORQUE GAIN(+) LOSS(-)
2000	266	101	89	− 4
2250	270	116	89	− 1
2500	269	128	91	− 10
2750	282	148	92	− 6
3000	295	169	94	− 5
3250	310	192	96	+ 3
3500	314	209	98*	+ 7
3750	315*	225	97	+ 13
4000	314	239	97	+ 14
4250	313	253	98	+ 22
4500	307	263	96	+ 25
4750	301	272	96	+ 26
5000	290	276	94	+ 27
5250	280	280	93	+ 27 (27 HP)
5500	269	282*	91	+ 31 (32 HP)
5750	252	276	88	+ 31 (34 HP)
6000	236	270	87	+ 31 (35 HP)*

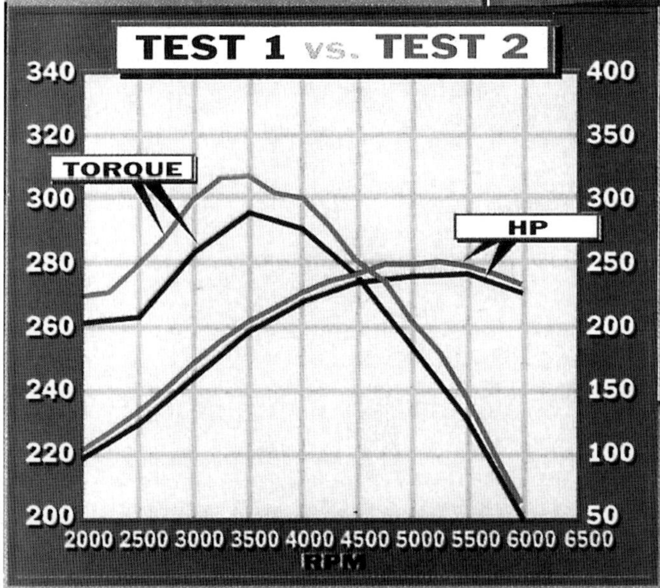

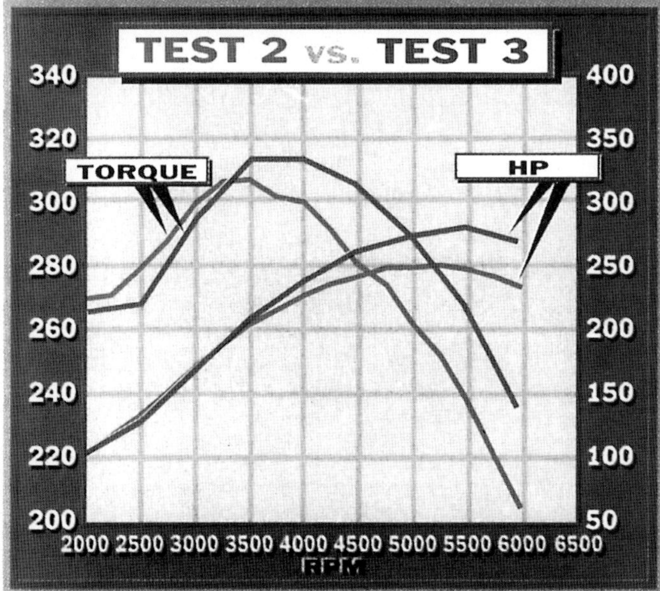

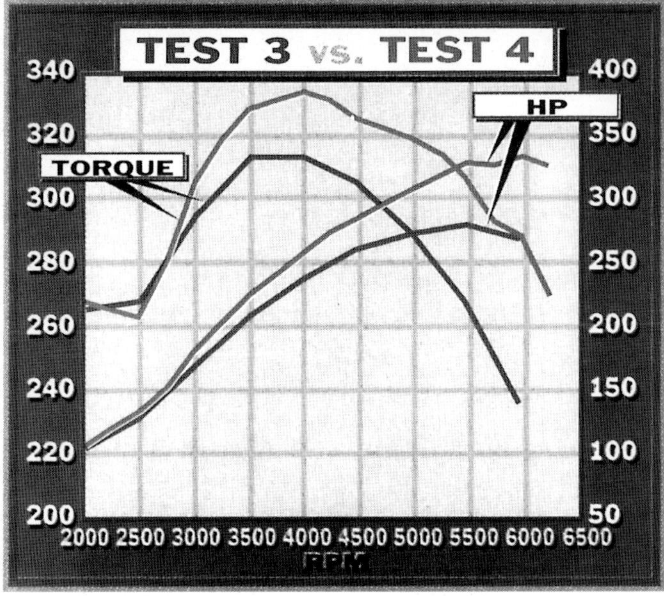

The first change was to outfit the RHO engine with a set of "real" 1⅝-inch headers from Hedman. They were installed on the dyno in "boat" fashion in order to clear the dyno stand. The longer-tube headers in combination with the improved exhaust ports produced dramatic improvements, especially in low-speed torque.

were in the above-3000-rpm range, since the larger RPM intake lost small amounts of power below 3250 rpm. From 3250 rpm on up, the torque increased *at least* 17 lbs.-ft., with the largest increase at 5500 rpm, where the RHO engine with intake and headers cranked out a whopping 38 lbs.-ft. more torque and *40 horsepower* over the stock configuration. Looking at just the intake change alone (not including the headers), the Edelbrock RPM intake was worth 34 horsepower over the stock intake and header combination at 5750 rpm! With just an intake and header swap, the RHO engine cranked out an impressive 282 horsepower at 5500 rpm and 315 lbs.-ft. of torque at 3750 rpm!

Just as we were about to yank the engine off the dyno, Bill Mitchell sent us a set of his latest Dart II iron heads for the Windsor Ford. With a stout 195cc intake port, 2.02-/1.60-inch valve diameters, and a decent exhaust-to-intake-port flow relationship, the heads appeared to be worth some power. There was some concern, however, that the large intake port would create a slight power loss below the torque peak. But once the heads were installed, the numbers told an amazing story.

The new Dart II iron heads require dedicated small-block Ford 1.6:1 rocker arms, which Kaufmann supplied. The RHO cam also necessitated a valve-spring change, since the Dart II heads are supplied with stock-type small-block Chevy valvesprings, which would never have cut it with the Ford roller cam. Bill Mitchell supplied a set of stronger valvesprings and retainers, and once they were installed, we were ready to test.

Again, af-

TEST 4

The same as the previous test with the addition of Dart II iron cylinder heads and 1.6:1 roller rocker arms. The heads added an approximate 0.5:1 compression ratio increase.

RPM	TORQUE (LBS.-FT.)	HORSEPOWER	VOLUMETRIC EFFICIENCY (PERCENTAGE)	TORQUE GAIN(+) LOSS(-)
2000	268	102	89	+ 2
2250	265	114	90	– 5
2500	263	125	90	– 6
2750	281	147	93	– 1
3000	306	175	96	+ 11
3250	319	197	97	+ 9
3500	328	219	99	+ 14
3750	331	237	100	+ 16
4000	334*	255	101*	+ 20
4250	331	268	101	+ 18
4500	326	279	100	+ 19
4750	322	291	100	+ 21
5000	319	304	99	+ 29
5250	315	315	99	+ 35 (35 HP)
5500	307	321	98	+ 38 (39 HP)
5750	294	321	96	+ 42 (45 HP)
6000	288	330*	96	+ 52 (60 HP)*

Our second modification was improving the induction side with the addition of an Edelbrock RPM dual-plane intake manifold. Compared with the stock Ford intake port, the RPM intake offers a substantially larger intake port. This proved to be worthwhile, for the RPM intake generated a 35-horsepower increase, when combined with the headers. A Victor Jr. would also work well in this application if you were looking to increase high-rpm power. The RPM dual plane is our choice for all but the most serious street duty.

If the B-303 roller cam that comes in the RHO engine is a bit lumpy for your taste, Kaufmann Products offers a dual-pattern roller cam with a shorter 214-duration intake profile. It provides 224 degrees of exhaust duration at .050-inch tappet lift and a 114-degree lobe-separation angle, which generates a much smoother idle. Power is still a strong point, since the cam offers much taller .520-/.544-inch valve-lift specs.

The RHO engine comes with forged, four-eyebrow pistons. When using the Dart II iron heads with a high-lift camshaft, you will probably have to remachine the valve relief pockets in the pistons, as the larger valves in the Dart heads could hit the pistons. Always check the valve-to-piston clearance when changing either the heads or cam.

ter the requisite tuning process, our first full pull on the RHO engine with the Dart II heads resulted in an amazing *60-horsepower increase* over the previous combination. The Dart II heads lost only six lbs.-ft. of torque at 2500 rpm and gained no less than nine lbs.-ft. of torque from 3000 rpm up to 6000 rpm! With this combination, the RHO made maximums of 330 horsepower at 6000 rpm and 334 lbs.-ft. of torque at 4000 rpm, increasing the airflow through the engine by well over 40 cfm at 6000 rpm! Even at the RHO head's peak horsepower point, the Dart II heads cranked out 39 more horsepower. Keep in mind, too, that the Dart II heads enjoyed the advantage of roughly half a point of compression increase over the RHO heads, which is worth some power. To be fair, we should mention that by using the 1.6:1 roller rockers on the Dart II heads, some power was no doubt gained over the stock, stamped rockers used with the RHO heads.

As you can see from this test, the RHO small-block Ford responded to every modification we cranked into it. Remember, this test was performed with a complete muffled exhaust system and pump gasoline, which means these numbers represent what you could expect from this engine installed in a car. Even more power could be obtained by pocket porting the heads, experimenting with camshafts, and choosing a more efficient muffler such as the Borla. Perhaps the best part of this entire engine deal is the price range; at this writing, the RHO engine costs approximately $1800. Regardless of whether you like your small-block mild or wild, the Ford RHO 5.0 engine is a great deal on a great deal of engine. **HR**

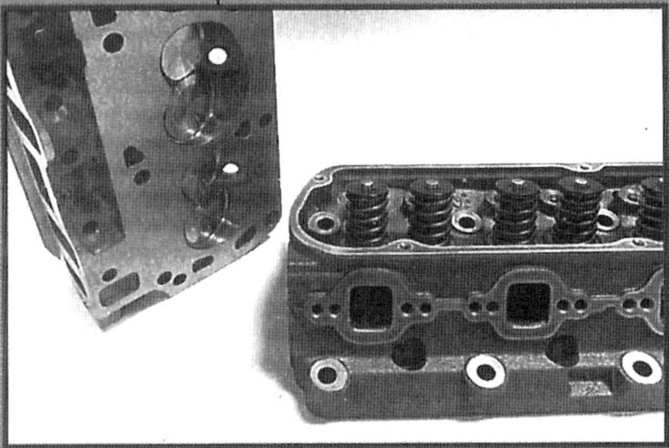

The Dart II iron Ford heads come completely assembled, or you can buy them bare. The chamber size is 64 cc's, the valves are 2.02/1.60, and the exhaust port is drilled for both exhaust-flange patterns. A little bowl-work on the exhaust side and some intake-port work could generate even more flow potential from these heads.

The Dart II heads come stock with low-po valvesprings that are fine for basic flat-tappet cams. But to control the RHO engine's radical B-303 roller cam, Bill Mitchell supplied larger valvesprings. The Dart II heads come with 3/8-inch screw-in studs and guideplates. You should use rocker arms designed for the small-block Ford (small-block Chevy rockers won't work), such as those offered by Crane, Edelbrock, or Competition Cams. The stock Ford stamped rockers deflect horrendously, especially when used with roller-cam valvesprings.

SOURCES

B&B Performance
Dept. HR
23190 Del Lago Dr.
Laguna Hills, CA 92653
(714) 586-0561
Underdrive V-belt and serpentine belt pulleys

Edelbrock Corp.
Dept. HR
2700 California St.
Torrance, CA 90503
(213) 781-2222
Intake manifolds, valvetrain

Fel-Pro, Inc.
Dept. HR
7540 N. McCormick Blvd.
Skokie, IL 60076
(708) 674-7700
Performance gaskets

Ford Motor Company—SVO
Dept. HR
17000 Southfield Rd.
Allen Park, MI 48101
(313) 845-2273
Ford high-performance equipment

Hedman Hedders
Dept. HR
P.O. Box 2126
Culver City, CA 90231
(213) 839-7581
Headers and exhaust components

Holley Performance Parts
Dept. HR
11955 E. 9 Mile Rd.
Warren, MI 48089
(313) 497-4000
Carburetors, intake manifolds

Kaufmann Products, Inc.
Dept. HR
12400 Benedict Ave.
Downey, CA 90242
(213) 803-5531
Late-model Ford performance products

McFarland & Associates
Dept. HR
390 Amapola Suite 1
Torrance, CA 90501
(213) 533-6100
Dyno testing, research and development

World Products, Inc.
Dept. HR
517 Orchard View
Royal Oak, MI 48073
(313) 583-1818
Dart II performance iron cylinder heads

By Marlan Davis

MOVE OVER CHEVY, HERE COMES FORD'S NEW 385-HP CRATE MOTOR

WINDSOR Dynasty

In the high-stakes game of competitive small-block crate engines, Ford Motorsport has raised the stakes of the game. It has matched the Bow-Tie boys' 345-hp ante—and raised it by 40! Later this fall, Ford will introduce 385-hp, 378-lbs.-ft.-of-torque, 351W crate motors that are complete and ready to drop into your favorite FoMoCo product. The engines will meet Ford's standard durability dyno torture test: 50 hours running at peak torque, followed by 50 hours running at peak horsepower! Each individual engine will actually be test fired before shipment. This represents a real milestone for Ford, which hasn't offered an engine nearly this powerful since the demise of the big-block Cobra Jets and 351-Boss motors back in the early Seventies. It took a lot of blood, sweat, toil, and tears to bring this motor to the general public. Let's take a closer look at the making of the newest "King of the Windsors."

Ford's 351W crate motor will be delivered complete from oil pan to intake manifold; from water pump and harmonic balancer to flywheel. All the customer needs to supply is a carburetor and an exhaust system. The engine uses modified Motorsport aluminum cylinder heads that are bolted to an F1JE-6009-FA Marine short-block assembly. These Marine assemblies consist of a 2-bolt main-cylinder block; brass freeze plugs; a cast-iron crank; 10:1 pistons; and a heavy-duty Marine/truck connecting rod assembly. Its performance-style piston-ring package utilizes a 1/16-inch moly-faced top ring, a 1/16-inch chrome-plated second, and a 3/16-inch standard tension, expanded rail-type oil ring. A forged crank and a 4-bolt-main block aren't included because they would require expensive full-race SVO components, and the cost is simply not justifiable in view of the existing off-the-shelf components that pass Ford's durability test.

The short-block (less cam) is based on Ford's proven Marine 351W engine. The hypereutectic pistons combine with Motorsport's Fel-Pro M-6051-A331 head gasket (8.8cc compressed volume) and 52cc combustion chambers to yield a nominal 10:1 compression ratio.

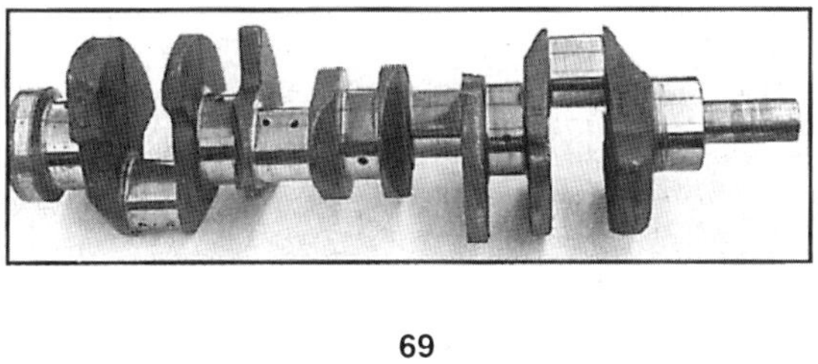

Like the rest of the engine, the factory-made cast nodular iron crank survived Ford's full-power durability torture test.

Considerable research went into the development of a new flat-tappet hydraulic camshaft for the crate motor. Ford's goal was to significantly increase power without beating up the valvetrain. The result is a stout dual-pattern grind (M-6250-A351), featuring .520-inch intake/.538-inch exhaust lift; 296 intake/306 exhaust degrees advertised duration (236 intake/246 exhaust degrees duration at .050-inch tappet lift); 109 intake/119 exhaust degrees centerlines; and 114 degrees lobe separation. Because this cam is designed for maximum horsepower, only eight inches of vacuum is present at idle. For street use, a Cam Dynamics or Summit Racing vacuum reserve canister is recommended. Automatic trans applications require a minimum 2500-rpm-stall-speed converter.

The cam's lobes activate anti-

The pistons are attached to heavy-duty Marine/truck 351W rods with pressed pins. These rods feature heavy-duty 3/8-inch diameter bolts and spot-faced bolt seats.

The cam is retarded two degrees and installed by using the nine-position multi-keyway double-roller timing chain (M-6268-A302). Tests have shown that this configuration produces the most power.

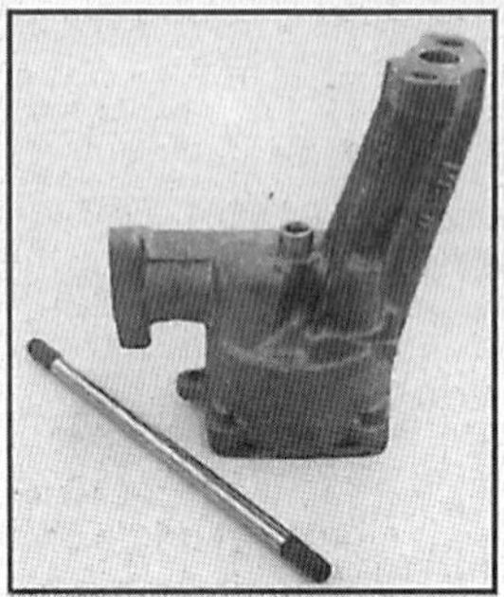

An M-6600-B3 high-volume oil pump and an M-6605-A341 heavy-duty 3/8-inch o.d. chrome-moly drive shaft ensure bulletproof lubrication.

used in the crate motor, the exhaust valves will have flow-enhancing undercut stems.

Edelbrock's new 351W Victor Jr. single-plane intake manifold was the induction system of choice. A Ford electronic distributor; a 157-tooth manual trans flywheel drilled to

pump-up hydraulic lifters (M-6500-A301), hardened 5/16-inch-diameter pushrods, and roller-tip, needle-bearing-fulcrum roller rocker arms to actuate the 1.94-inch intake/1.60-inch exhaust valves, which are installed in the Motorsport aluminum heads. These valve sizes are the largest possible diameter compatible with off-the-shelf pistons and a short-block.

The crate motor's cylinder heads feature minor porting underneath the valve seat in the short-turn radius area. Presently, Motorsport sells bare, unported aluminum heads machined for 1.94/1.60 valves under part No. M-6049-J302. Fully assembled aluminum heads are also available under M-6049-K302. The new "K" assemblies will have larger 2.02-inch intake valves; and unlike those

351W ENGINE INSTALLATION NOTES

Ford's 351W block is one-inch taller and 1.1 inches wider than 289/302 Ford small-blocks. The late-production block used to build the new crate motor also has several alterations compared with Sixties- to Seventies-vintage 351Ws that preclude total interchange with original 351W equipped cars. Here are some of the changes and differences and their solutions.

FRONT COVER & FUEL PUMP

There is no provision for mounting a mechanical fuel pump. If you presently run a mechanical pump, either substitute an earlier 351W front cover with mechanical fuel-pump provisions or convert to an aftermarket inline electric fuel pump. A mechanical fuel pump eccentric (D0AZ-6287, outer; D5AZ-6287-B, inner) must also be installed on the front of the camshaft.

HARMONIC DAMPER

The crate motor's NHRA-approved balancer is compatible with '70-up four-bolt crank pulleys. A spacer is

V-BELT-COMPATIBLE FRONT-COVER COMBINATIONS

The timing cover, timing pointer, water pump, and damper must be compatible. The crate motor comes set up for serpentine belt applications. For V-belts, the following combinations are compatible. Many times, existing 289/302 components will also bolt on.

INTENDED USE	COVER	POINTER	DAMPER	WATER PUMP
'69 and earlier, three-bolt crank pulley, passenger's-side water pump inlet	E7AZ-6019-A[1]	C8AZ-6023-B[2]	C9OZ-6316-A[2]	D3UZ-8501-A
'70-up, four-bolt crank pulley, driver's-side water pump inlet	E7AZ-6019-A[1]	C9TZ-6023-A	Use existing damper	D6ZZ-8501-B

1. Grind off crank trigger mounting pad, as required for timing pointer clearance. Also, use this part with serpentine belt drives when running a mechanical fuel pump.
2. Obsolete; procure from salvage yard.

PART NO.	AR[1]	COMPONENT
M-6505-B302	4	Exhaust valve—1.60" dia.
M-6507-A302	4	Intake valve—1.94" dia.
Crane #998932	8[2]	Valvespring—dual
M-6214-A50	8	Retainer—valvespring
M-6518-A351	8	Keeper—valvespring
AR 3162	8[2]	Seat—valvespring
M-6564-AS51	4[3]	Roller rocker arm—1.60:1
M6565-B332	8	Pushrod—.312"x8.152"
M-6571-A221	8	Valve-stem seal (PC type)
M-6566-B302	4	Pushrod guideplate
M-6527-B311	8	Rocker arm stud

1. Amount required per cylinder head.
2. Ford Motorsport part No. not yet assigned.
3. Sold in unit packages of two, so each head requires four packages. A complete engine set of 16 rockers is available under M-6564-A351.

351W CRATE MOTOR DYNO TEST*

ENGINE SPEED	TORQUE (lbs.-ft.)	HP
3500	331.6	221.0
3750	346.6	247.5
4000	363.5	276.8
4250	372.6	301.5
4500	377.8	323.7
4750	370.6	335.2
5000	368.4	350.7
5250	360.6	360.5
5500	353.4	370.1
5750	351.4	384.7
6000	328.2	374.9

*Results of step test conducted by ECS/Roush on Ford Motorsport 351W developmental crate motor prototype.

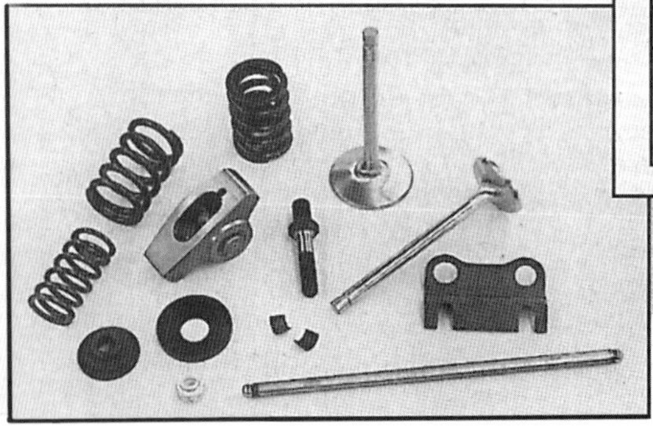

New heavy-duty valvetrain components are being introduced on this motor to specifically complement the short-block, cylinder heads, and camshaft. These parts will also be offered separately.

Since all current stock Ford engines are fuel injected, the front-cover assembly no longer has mechanical fuel-pump mounting-bolt provisions. The late truck aluminum reverse-rotation water pump (E7TZ-8501-A) is designed for serpentine belt applications (its inlet is on the driver's side).

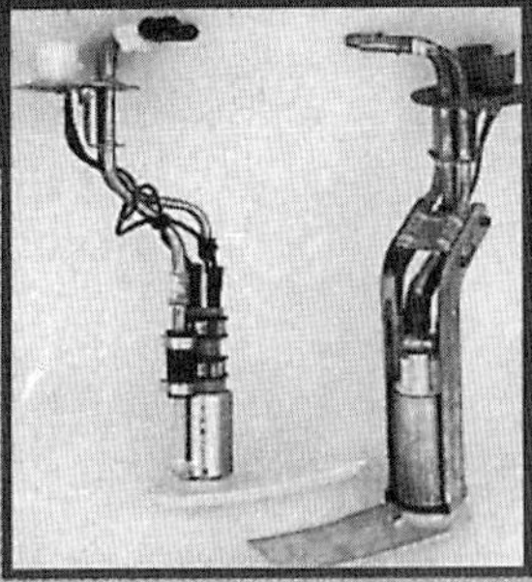

When switching over from fuel-injected Mustangs to carburetion, a six psi low-pressure '84-'86 SVO Mustang turbo 4-cylinder in-tank electric fuel pump (E3ZZ-9350-A) bolts in place of the existing in-tank high-pressure unit.

Rick Roback of Union Lake, Michigan, installed one of the first 351W crate motors into his '85 5.0L Mustang. With the Motorsport heavy-duty T5 5-speed and 8.8-inch differential equipped with 4.56:1 gears and an Airlocker centersection, the car so far has run 11.68 seconds at 115 mph on slicks! He used the 1⅝-inch-tube 351W header set (M-9430-J351) to install the engine. For use with a Lakewood scattershield, the right-side header had to be extensively modified and the crossmember slightly notched. The scattershield also required trimming.

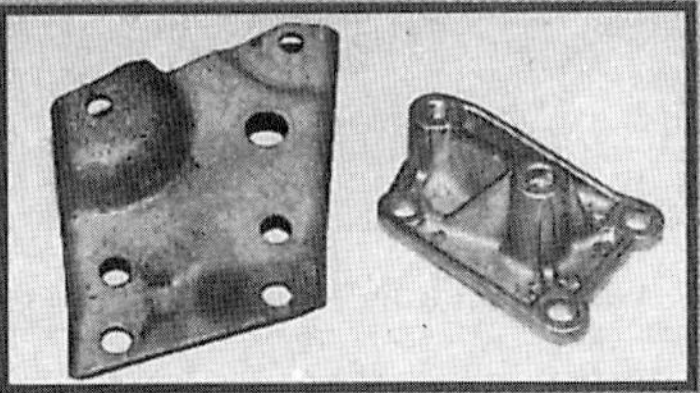

Modified air-conditioning and power-steering brackets are required when using the 351W in late Mustangs or other Fox-bodied cars. Presently, Total Performance can modify your existing brackets for $75; they also offer 351W air-conditioner-elimination kits.

The motor's degreed crankshaft damper (M-6316-C351) has the correct 351W unbalance weight. When serpentine-belt-drive pulleys are used, a .630-inch-thick spacer (M-6359-A351) must be purchased separately to achieve proper spacing.

The crate motor's oil pan (shown) is compatible with late Mustangs and other Fox-bodied cars. Earlier cars need a front-sump 351W pan (D5OZ-6675-A) and a pickup (D7AZ-6622-A).

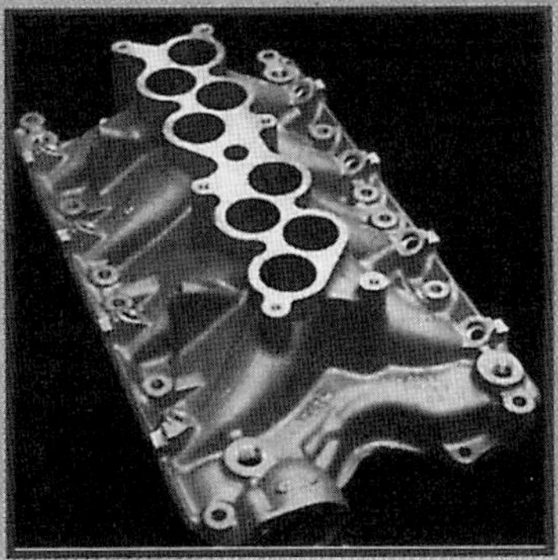

Ford Motorsport now offers this high-flow GT-40-style, EFI-compatible 351W lower intake manifold. When used in conjunction with other Motorsport high-flow EFI parts, it allows use of the new crate motor with late-passenger-car fuel-injection systems.

required with late serpentine-belt- drive setups. You'll need a '69 351W damper for '69-and-earlier three-bolt crank pulley applications. Original 3-bolt combos also use a different water pump and front cover (see chart).

WATER PUMP

The reverse-rotation aluminum pump is used with serpentine drive belts and is incompatible with earlier V-belt setups (in fact, the late pump's pulley bolt pattern is different). If you're using V-belts, a corresponding earlier pump with the correct-facing inlet is required (see chart).

ACCESSORY DRIVES

Due to the 351's higher deck height, 351 brackets are usually needed when changing over from a 289 or 302. The problem occurs on late-model passenger-car serpentine-belt-drive applications where there is no corresponding 351W application. On these models, the existing 302 air-conditioning and power-steering brackets must be modified to fit the 351W. Eventually, Ford Motorsport will offer the needed brackets.

EXHAUST

In many instances, existing stock 289/302 iron exhaust manifolds will work. If you want to run headers, a dedicated 351W set is usually required.

CLUTCH LINKAGE

The late-model 351W block no longer has a clutch-shaft-pivot-ball hole. This isn't a problem on late Mustangs or Fox-bodied cars, which use cable-actuated clutches, but it will affect retrofit in earlier chassis that use a bell-crank-style mechanical clutch linkage; the user will either need to purchase a pivot-ball mounting bracket from Total Performance; convert to hydraulic linkage by using a McLeod or Tilton integral hydraulic throwout bearing; or purchase Kaufmann Products' cable clutch linkage conversion kit.

FLYWHEEL & FLEXPLATE

All 351Ws use flywheels with 28.2 ounces of unbalance, the same as most '80 and earlier 289/302 engines. Hence, any early flywheel or flexplate with 28.2 ounces of unbalance can be used. All '81 and later 302s use a 50-ounce unbalance factor and cannot be used on a 351W, without rebalancing the flywheel or flexplate. Original 351W production flywheels and flexplates were all 164 tooth. The crate motor's steel flywheel has 157 teeth, the same as most 289/302 manual trans applications; the bellhousing must match the flywheel size to properly mount the starter. The C6 flexplates cannot be used on C4/C5/AOD applications, and vice versa.

EFI COMPATIBILITY

Those desiring to utilize the crate motor while retaining the existing passenger-car factory fuel-injection system will need a truck 351W EFI distributor (E7TZ-12127-D); a new Motorsport GT-40 351W lower intake manifold (M-9K461-A51); a high-flow upper intake manifold (M-9424-A51); a high-flow throttle body (M-9E926-A302); a mass air conversion kit (M-9000-A51, automatic trans, or M-9000-B50, manual trans—except '88 California and all '90-'91 5.0L GT Mustangs, which came stock with a Mass Air Flow sensor); a high-flow fuel pump (M-9A407-A50); new 24 pound/hour fuel injectors (M-9F593-A302); and other miscellaneous components as listed in the current Ford Motorsport catalog.

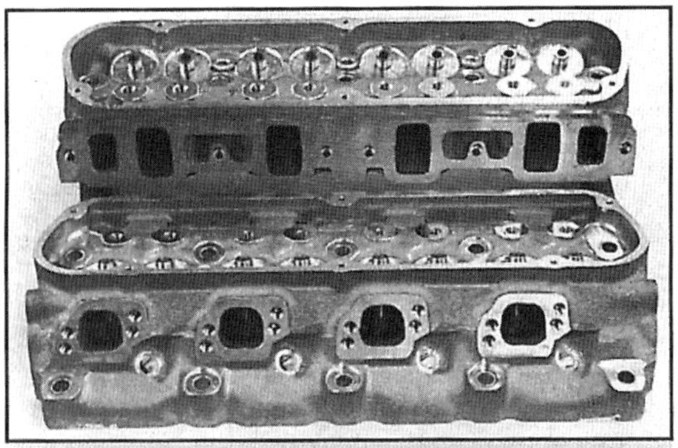

Cast from 356-T6 aluminum alloy, the aluminum heads' 60cc runner volume exhaust ports exit 1/16 of an inch higher than stock. However, the exhaust manifold's dual mounting-bolt pattern allows them to accept stock exhaust manifolds or street headers, as well as race headers with up to 1¾-inch primaries. The 170cc intake runners have no provisions for exhaust crossover heat or EGR. Adjustable, stud-mounted rocker arms are required.

A 351W electronic ignition distributor (E2AZ-12127-E) with centrifugal and vacuum advance and an M-12029-A302 coil is standard, but you must supply the control module. Either the stock Duraspark II, Motorsport "extra-performance," or full-race "Ultra High Energy" (MSD-style) system is acceptable.

The 52cc combustion chambers are machined for 1.94-inch intake/1.60-inch exhaust valves. The sharp edges in the exhaust port bowl area are ground out to produce a smooth transition from the valve seat top angle through the short-turn radius. Three-angle valve seats are ground at 45-degrees, with a 30-degree top angle and 55-degree bottom angle.

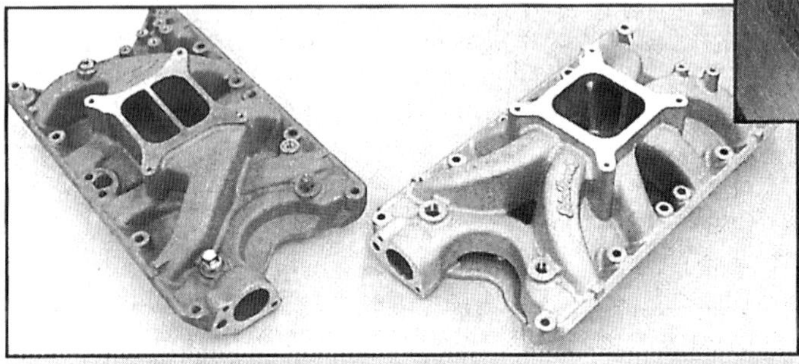

The initial production crate motors are designed for maximum top-end horsepower, so they'll come with Edelbrock's new 351W Victor Jr. single-plane intake (Edelbrock No. 2981; Ford-Motorsport No. M-9424-V351, right). A later midrange torque package will feature the Motorsport dual-plane intake (M-9424-A341, left).

accept either early, long- or late-diaphragm-style clutches; an NHRA-approved damper; and an aluminum water pump complete the crate-motor assembly, which will carry Ford Motorsport part No. M-6007-A351.

Tentative Motorsport plans eventually call for a less-expensive version of the crate motor, which will be available with the Motorsport dual-plane intake manifold and (M-6049-L302) cast-iron GT-40 heads. With their smaller ports and reduced 1.84/1.54 valve sizes, crate engines with GT-40 heads will probably be down about 20 hp from the aluminum-headed assemblies, although midrange torque should come up, thanks to the dual-plane intake manifold. The complete engine assembly with cast-iron heads has been temporarily assigned part No. M-6007-B351. We'd recommend the dual-plane engine assembly for heavy cars with automatic transmissions. Apparently, Motorsport agrees, because the projected cast-iron assembly is slated to receive an auto-trans flexplate in lieu of a flywheel and may also have a slightly milder cam. While no official list price has been established as we go to press, Ford is planning to sell the aluminum crate motors for "under $4000." That's more expensive than the competition's suggested $3200 list price, but you also get nearly 12 percent more horsepower.

The original dyno development work was conducted by ECS/Roush using the Victor Jr. combination, which was fed by a 780-cfm Holley Model 0-4628 vacuum secondary carburetor. As originally installed on '70 429-SCJ auto-trans applications, the carb came with No. 70 jets in the primaries and 83s in the secondaries. In order to equalize exhaust-gas-temperature thermocouple readings on the test engine (thereby achieving optimum air/fuel mixture distribution), it proved necessary to stagger-jet the carb as follows: left front, No. 70; right front, No. 69; left rear, No. 78; right rear, No. 77. Exhaust chores were handled by a non-optimized, well-used set of dyno headers with approximately 1⅝-inchx40-inch primary tubes that dump into 3½x8-inch collectors.

The tests were run on Sunoco 94 octane unleaded gas, with 36 degrees of total timing checked at 5000 rpm. (A Ford spokesperson says the engine should run satisfactorily on 92 octane or better gas.) For the extended dyno pulls, ultra-cold Motorcraft AG901 spark plugs gapped at .040-inch were required (use AG32 or AGS32 in the vehicle). In this dyno test configuration, the engine produced 385 hp at 5750 rpm and 378 lbs.-ft. of torque at 4500 rpm. Over 300 hp was produced from 4250 rpm all the way through 6000 rpm, the highest rpm tested. Torque output exceeded 350 lbs.-ft. from 4000 through 5750 rpm (see chart). This engine really oughta fly!

Of course, dynos don't win races. How will this combination perform in the real world? Early next year, HRM plans a high-noon face-off between Chevy's 350-H.O. motor installed in a Camaro and the new Windsor 351 crate motor installed in a Mustang. Will the mighty Mouse dethrone the Windsor dynasty, or will the King's Ford blue-blood prevail? Stay tuned. **HR**

SOURCES

ECS/Roush
Dept. HR
12447 Levan Rd.
Livonia, MI 48150
(313) 591-4384

Ford Motorsport Performance Equipment Ford Motor Co.—SVO
Dept. HR
17000 Southfield Rd.
Allen Park, MI 48101
(313) 337-1356

Total Performance
Dept. HR
44050 N.
Groesbeck Hwy.
Mt. Clemens, MI 48043
(313) 468-3673

Kaufmann Products
Dept. HR
12400 Benedict Ave.
Downey, CA 90242
(213) 803-5531

Whittier Lincoln-Mercury
Dept. HR
13003 Whittier Blv
Whittier, CA 9060
(213) 698-5635

HOT PURSUIT!

HOW FORD MAKES THE MUSTANG FLY

While the organized lunacy we call government continues to restrict and legislate car nuts at every level, Ford is quietly cranking out full-blown street hot rods just as fast as they can screw them together. The '85 Mustang GT carries enough high-performance ordnance to level EPA headquarters and promote mass hysteria among the disciples of doom. Now here's a car that truly has the right stuff—a musclecar *extraordinaire*. Sure, some of the old musclecars came with Holley carburetors, but how many of them ever turned a tire with a roller camshaft and honest-to-God tube headers actually installed in the car? Virtually every component in the HO engine package has been examined from the hot rodder's critical viewpoint, and the result is the undeniable conclusion that efficiency, high performance, and clean air go hand in hand. We're talking about a smog-legal, 14-second, 130 mph Mustang that gets better fuel economy than any performance Mustang ever built. Ford has beat the feds at their own game, and the only reason they haven't been penalized is because they played by the rules. That's why the bad guys rewrite the rules every few years.

The men responsible for this performance are Jim Clarke and his cadre of hot rodders down the hall at Ford's Small V8 and V6 Development. They like to call it the engineering department, but hey, we know what they're up to in there, and it ain't buildin' motors for farm implements. They've taken a six-year old car and a basic complement of hot rod components and run right up beside all those high-tech, fuel-injected, computer-controlled supercars. The '84 Mustang HO engine has been hot rodded in the finest traditional manner, making it more than capable of meeting the modern day performance challenge. With their high-performance mindset and sterling record of achievement, it's not unreasonable to speculate that these are the men who might just deliver the first domestic supercharged, four-valve V8—and sooner than you think. While other engineering departments just roll over and flap on incessantly about the new era of 4-cylinder and front-wheel-drive performance, Ford's performance people are stepping up with the heavy artillery. No question about it—they're our kind of guys. God save the heroes!

Ford's high-performance recipe for '85 includes a traditional setting of headers, Holley carburetor, roller camshaft, and dual inlet high-flow air cleaner.

Stainless steel tube headers originally designed for the '83 Mustang have been updated with improved carburetor air heat source and larger exhaust system.

The 585-cfm 4-barrel carburetor is built by Holley Carburetor to Ford's exclusive specifications. It is tailored specifically to the total performance package.

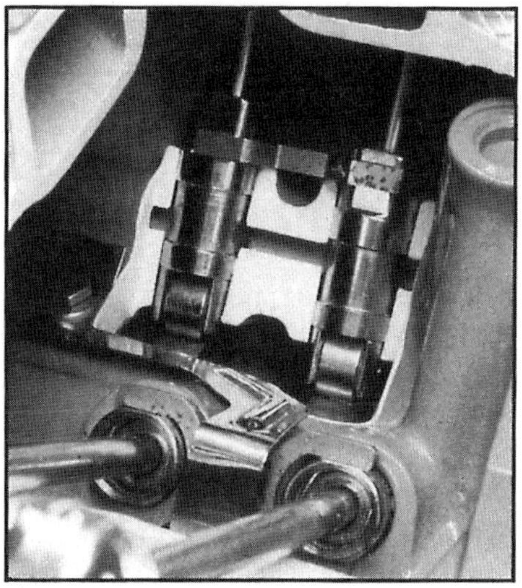

Ford's trick hydraulic roller cam features .445-inch valve lift, 266 degrees advertised duration, 244 degrees at .050-inch lift and 115-degree lobe centers. Performance is smooth and hot.

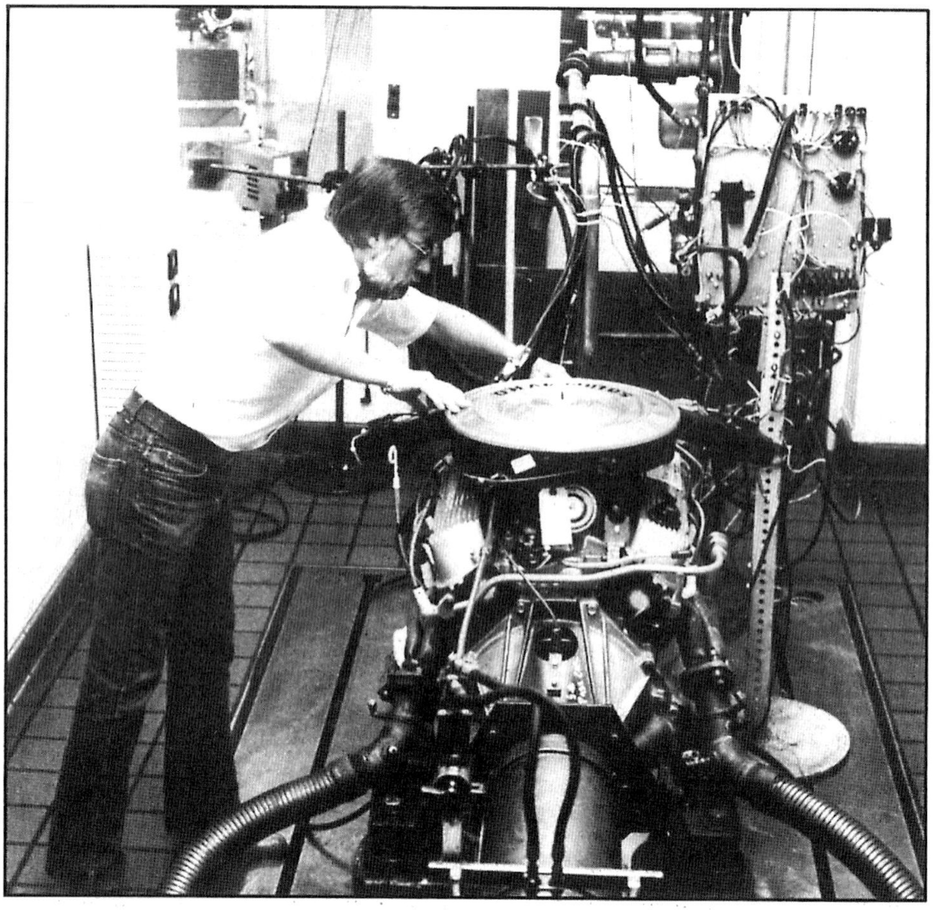

SPEC SHEET

1985 MUSTANG GT

ENGINE:

Type	OHV 90-degree V8, HO option
Bore & Stroke	4.00 x 3.00-inch
Displacement	5.0L, 302-cid
Compression Ratio	8.4:1
Induction System	585-cfm Holley 4V
Exhaust System	Stainless steel tubular headers
Camshaft	Hydraulic roller tappet design
Horsepower	210 @ 4400 rpm
Torque	270 @ 3200 rpm

DRIVETRAIN:

Transmission	5-speed manual Borg-Warner T-5
Final Drive Ratio	2.73:1 with 3.08:1 opt.

SUSPENSION:

Front	Hybrid McPherson strut with springs mounted on lower control arms, 1.30-inch stabilizer bar
Rear	Four-bar link coil spring on lower arm, .670-inch stabilizer bar, Quadra Shock system with vertical shocks, and Freon bag hydraulic horizontal axle dampers
Wheels	15x7 cast aluminum
Tires	Goodyear Eagle VR60, P225/60VR-15
Brakes	10.06-inch front discs, rear drums with vacuum assist

GENERAL:

Curb Weight	3130 lbs.
Weight Distribution	56.15, front; 43.85, rear
Fuel Capacity	15.4 gallons
Fuel Economy	Observed, 15 mpg city, 22.5 mpg highway

PERFORMANCE:

0 to 60	6.5 seconds
Quarter-mile	14.90 @ 92 mph
Lateral Acceleration	.80g
Top Speed	133 mph

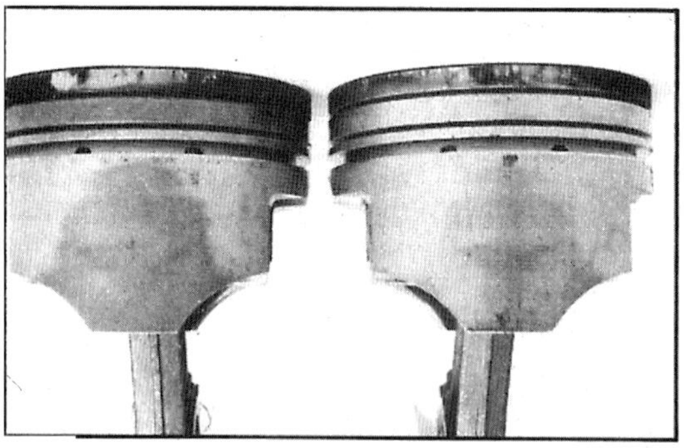

Comparison of high-mileage pistons shows reduced wear pattern on the new design (right). Friction reduction was also accomplished by tightening up the skirt loading area and removing stress from the outer portion of the skirt.

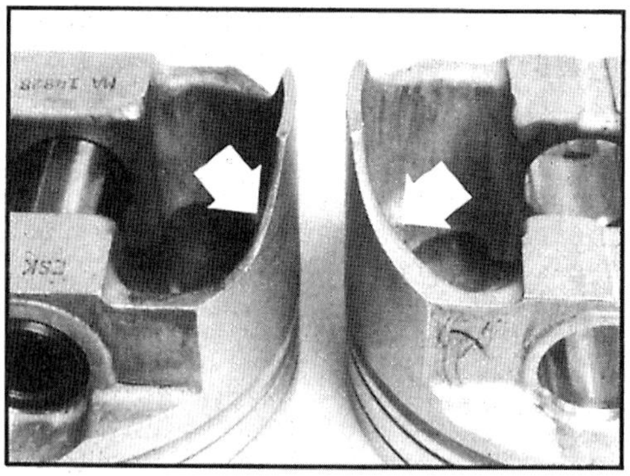

Redesigned and strengthened piston skirt shown on the right has proven totally reliable in service. Fix included reshaping the cam and barrel face profile, while adding material for strength.

THE ANXIETY & THE ECSTASY: HOW THEY MAKE IT LIVE

A major complaint among Ford enthusiasts was the unexpected delay in the delivery of the potent 210 hp HO engine package, a powerplant originally expected to debut near the end of 1983. Some even suggested that Ford's explanation regarding a problem with piston durability was a smokescreen designed to conceal their lack of concern, and that they should be taken to task for advertising a product that was unavailable. We say . . . so what! Are we really expected to bite the hand that feeds us? Perhaps the on-schedule delivery of a 3-cylinder Dozer GT would prompt them to part with their mad money in the name of expediency. We think Ford deserves the benefit of the doubt. These guys may not be pushing the outside of the envelope with modern technology, but let's face it: at least they're keeping the walls from caving in around us.

Of course there are all sorts of reasons for a delay of this nature, and unexpected development problems are nothing new in the auto industry. Commendably, Ford opted to go with an extended development program when 50K durability testing revealed small skirt cracks in the forged piston design they had resurrected for the HO engine. A new piston design was implemented and an improved ring package was evaluated at the same time. We say they did the right thing, and we suspect that those who decry their efforts would also bleat the loudest while waiting in the warranty line. Anyway, it's all water under the bridge now. The real Mustang GT is a certified reality. No need to bring your spurs, cowboy. This pony likes to run. **HR**

POWERED by FORD

Ford is Back in Performance With a Whole New Heavy-Duty Parts Program

By Marlan Davis

For at least the past year we've been reporting that Ford is going back into the performance parts business, but until now actual physical manifestations in the form of tangible hardware were lacking. However, we're happy to report that the initial parts generated by the program—a redesigned Cleveland cylinder block and new aluminum heads—have recently made their first public appearance at Bud Moore Engineering (400 N. Fairview Ave., Spartanburg, SC 29302, 803/585-8155). Moore is one of the first Ford racers to receive these pre-production pieces, in order to test and fine tune them in preparation for their availability to the general public.

The Cleveland block is actually the renowned "Australian" casting that previously had been imported into this country on a limited "back door" basis for use by a lucky few Ford racers. However, the new 351C cylinder block is presently available through Ford's new parts network. Cast with a higher nickel content than standard Cleveland blocks, the 4-bolt main casting features thicker cylinder walls and beefed up main webbing and oil pan rails.

Though, good as the new casting is, it's only an interim solution. Eventually, a beefed-up Windsor casting that can be produced and machined on existing production tooling will replace it. The new Windsor castings will also feature beefed-up main webs and oil pan rails, but in addition they will have fully siamesed cylinder walls (allowing even larger overbores) and new 4-bolt main caps with splayed outer bolt holes that offer increased lower-end strength and support.

A replacement for the long-discontinued Boss 302 block will be the first of these new Windsor castings to appear, eventually followed by a higher deck 351W version for use in building 351 engines when the current supply of Cleveland Australian blocks is exhausted. This 351 Boss/Windsor block (for want of a better term) will use the same deck height as a production 351W, which is actually .300-inch higher than the current 351C block, enabling racers to run even longer connecting rods for additional top-end power. Instead of having the unique water pump housing extension on the Cleveland block, the eventual 351 replacement will accept Boss 302-style conventional covers, water pumps, and cylinder heads. But even though the external dimensions are those of a 351W, the block's main saddles will come machined to accept the 351-Cleveland crank with its 2.749-inch-diameter journals (as opposed to the 351W's 3.0-inch journals).

To bolt on top of these three new blocks (or any of the old 4-inch bore ones), Ford has a new aluminum cylinder head. Ford engineers started with a clean sheet of paper when designing these heads, incorporating all the advances discovered in 10-12 years since the original Boss heads were produced. The troublesome exhaust port was completely redesigned, giving way to a new high-efficiency, round-port design. Inlet port entry was also slightly raised; although the overall port volume was reduced by 12 percent, flow is reportedly increased by 15 percent, a significant gain in the intensely competitive world of cylinder head design. Ford reports the gains in the recontoured port resulted from a reduction of mixture turbulence.

Water passages were redesigned in an attempt to achieve heat rejection characteristics in the aluminum head comparable to the old cast iron head. A long straight-through water passage was also added beneath the exhaust port to provide needed additional cooling in that area. Provisions are made in the water jacketing so the head will easily accept drilling for the Boss 302-style water outlet passages, and a modification sketch is included with each head (eventually, two different part numbers will be offered: one drilled for the Cleveland's head-to-block outlets and the other for the Boss 302's head-to-inlet manifold water passages).

Insert valve seats are naturally required with the aluminum head, so to provide clearance for them the intake

POWERED by FORD

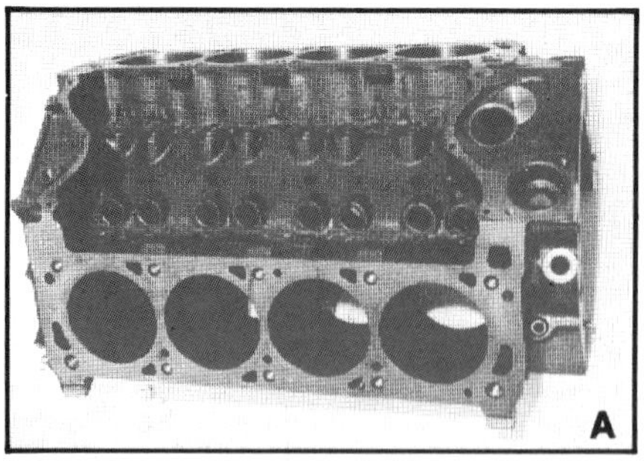

A

A *Up to .060-inch overbores are now feasible on the new thick-wall block, whose new "squared-off" cylinder wall core design is better able to withstand high-thrust loadings, thus virtually eliminating split walls. Each cylinder wall is now guaranteed to be at least .170-inch thick.*

B1

B2

B *Compared to standard Cleveland blocks **(1)** the new "Australian" block **(2)** has considerably more meat in critical bottom end areas, including filled-in pan rails and main webs.*

New Ford Heavy-Duty Service Parts

Part No.	Description
M-6010-A4	**Cylinder Block and Case,** Boss 302, beefed-up main webs, 4-bolt main bearing caps with splayed outer bolt holes, siamesed cylinder walls, high nickel content
M-6010-A331	**Cylinder Block and Case,** 351W deck height, machined for 2.749-inch 351C main crank journal diameter, 4-bolt main bearing caps with splayed outer bolt holes, siamesed cylinder walls, high nickel content (available when current supply of 351C castings is exhausted)
M-6015-A3	**Cylinder Block and Case,** 351-Cleveland ("Australian" casting), beefed-up main webs, thick cylinder walls, 4-bolt main bearing caps, high nickel content
M-6049-A3	**Cylinder Head,** 351-Cleveland/Boss, features redesigned intake and exhaust ports, revised water jackets, made from aluminum (comes with complete instructions containing water passage mods to enable use on Boss 302)
M-6051-A341	**Head Gasket,** 4-inch bore small-block engines, special composition for high-output endurance applications
M-6303-A4	**Crankshaft,** forged steel, Boss 302 3-inch stroke
M-6303-A341	**Crankshaft,** forged steel, 351-Cleveland 3.50-inch stroke
M-6316-A3	**Crankshaft Dampner,** heavy-duty Trans-Am style for Boss 302/351C, comes fully degreed
M-9424-A	**Intake Manifold,** 351C with new high-port aluminum head M-6049-A3
(Not assigned)	**Rocker Arm Cover,** Boss 302/351C, with integral rocker arm stud girdle

valve head diameter was slightly reduced from 2.19 to 2.16 inches. Still, the Ford valves are larger and heavier than those found in the competition's small-block, requiring the rocker arm stud bosses to be beefed up and the spring pockets deepened in order to run successfully with today's radical cam profiles. Additionally, a new revolutionary rocker arm cover with an integral rocker arm stud support girdle is in the works; it is said to positively end stud flex and rocker arm "shuffle" even with .650-plus-inch lift cams. And, since the intake port has been recontoured, Ford also plans to offer an all-new intake manifold.

But that's only the tip of the iceberg. Coming within the next six months are a new special compositon head gasket with special soft-wire inserts designed to satisfactorily seal the aluminum head to the 4-bolt/cylinder Cleveland block; Boss 302 and 351C forged cranks, mushroom cams, and tappets; and rereleased heavy-duty guide plates, harmonic balancers, nuts, bolts, and miscellaneous hardware.

A little further on down the road, expect to see new aluminum Boss 429 blocks and heads and high-performance 4-cylinder parts (including perhaps a new Indy motor). Unlike their old Muscle Parts program, Ford is planning to take a leaf out of their Brand X competitor's notebook and make these parts widely available to the general public, who will be able to purchase them not only at their friendly Ford dealers, but also through regular speed shops and special warehousers like Bud Moore. Ford doesn't intend to copy or duplicate aftermarket parts, but in cases where the speed industry is either unable or unwilling to do the job, Ford will step in. Just what is available right now from the aftermarket can be visualized by an overview of what goes into a typical Bud Moore NASCAR engine.

Moore's Power

Up to now, parts availability problems have plagued Ford racers, but loyal Dearborn devotees like Bud Moore and the Woods brothers have stuck it out, making due with what they had. Despite their dependence on old, scarce parts and a lack of good aftermarket replacement parts, these racers have managed to remain competitive.

When he could get them, Moore of course preferred the beefy Australian blocks. No matter which block is used, the crack-prone No. 2 and 4 main caps are replaced by stronger steel caps. Factory crank forgings have heretofore been rare, so a Moldex steel billet is used instead. Bud designed his pistons himself to work successfully with his modified head combustion chambers; usually the Venolia-produced slugs come .030-over. They are set up for a Speed-Pro 1/16-1/16-3/16 ring set, that features a plasma-moly top, cast iron second, and low-tension oil ring. Ring-end gap varies according to the track, being widened for the super speedways to prevent butting. Detonation is also a problem on the long ovals, so the compression is sometimes dropped by up to one full point from the 13:1 run on the shorter tracks.

Bud uses Precision Machine Shop piston pins. These longer-than-stock Ford pins are made from high-grade tool steel and are of taperwall configuration. Their OD is increased from the standard Ford .911-inch to the Chevy .927-inch size. Since Bud has had trouble with floating the pins on the Ford engine, he does away with pin locks

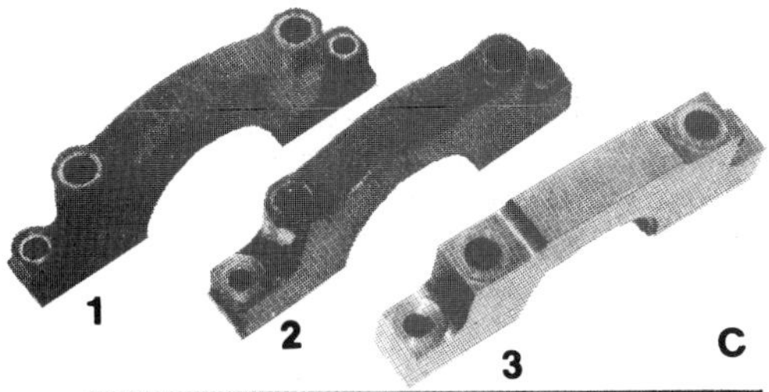

C *The new 4-bolt-main bearing cap (**2**) has more meat around the outer bolt holes than earlier cap (**1**). While it should be acceptable for drag racing and short-track use, for super speedway use Bud Moore still recommends replacement of the crack-prone No. 2 and 4 caps with Precision Machine Shop's special steel cap (**3**).*

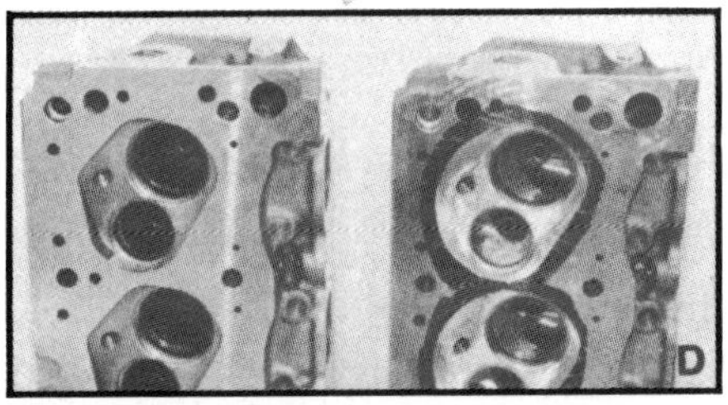

D *Since head porters have no consensus as to optimum combustion chamber configuration, Ford left the chambers on the new head virtually identical to the '71 and earlier Boss/4V configuration **(left)**. Moore has found that opening up the spark plug side of the head almost to the head gasket diameter enhances chamber efficiency on 351-sized engines when used in conjunction with a complementary piston dome. The area around the valves is also unshrouded.*

E *If anything, the Boss/Cleveland head ports are slightly too large. Moore modifies the original port **(1)** by adding material to the floor **(2)**. Now Ford does it for you **(3)**. So far, Moore's mods to the new heads have been limited to a general cleanup, polish, and gasket match **(4)**. Porter Greg Moore reports they flow at least as well out of the box as the best cast-iron modified ports they'd been able to come up with.*

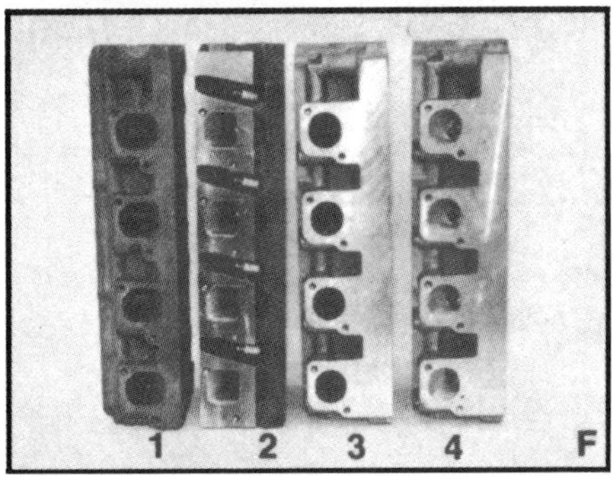

F *The exhaust ports have long been the Cleveland's Achilles' heel. To obtain decent flow, racers lopped off the original port **(1)** completely, replacing it with a fabricated plate **(2)** that featured completely recontoured ports with a higher exit. The new Ford head **(3)** incorporates the high exit but uses a round, instead of rectangular, cross-section. So far, Moore has been unable to achieve quite the same flow with the round ports as he did with his fabricated square port-plate, despite a polish and cleanup **(4)**.*

G *Since the Ford has larger valves than the competition, correspondingly higher rate springs are needed. The new head **(bottom)** features deeper spring pockets and beefed-up rocker stud bosses to accommodate the new generation of stiffer springs.*

Bud Moore NASCAR 351 Ford Specs

Clearance	
Main bearing	.0028-.003
Rod bearing	.0025-.0028
Rod pair side clearance	.018-.021
Piston-to-wall	.008-.010
Endplay (#3 is thrust bearing)	.006
Piston-to-valve @ lash	
Intake	.070
Exhaust	.115
Deck height	-.006 to -.010*
Gasket compressed thickness	.037
Combustion chamber volume (cc)	68
Ring end gap	
Short track	.018
Super speedway	.021-.022
Critical Torque (lbs./ft.)	
Head-to-block	120
Rods	100
Intake manifold-to-head	35
Main cap-to-block	95-100

* varies by track

and presses the pin into the piston end of the rod.

Carillo Grand National connecting rods that feature a beefed-up 100-gram-heavier rod cap go into Moore's motors. The rods are 6.0 inches in length (compared to the stock 5.780 inches), which Moore says work better on both the short and long tracks.

The big ends of these rods and the crank's rod journals accept TRW Clevite 77 NASCAR bearings designed for the 2.1-inch Chevy rod journal, since Moore believes this special bearing is the best obtainable. Standard 351C main journals are retained, however, and use regular CL77 bearings.

Engine block prep and assembly is straightforward, but meticulous. Naturally, the blocks are thoroughly deburred and the main bearing webs radiused. An 825-grit stone on a Sunnen CK10 hone is used to final hone the block with a torque plate installed. Unlike some other Cleveland racers, no radical oil system surgery is deemed necessary. Top end oiling is restricted both by offsetting the cam bearings' oil holes in relation to the block's cam journal oil holes and by reducing the depth of the tappet wafer washer groove.

The tappets themselves are of the mushroom type with a wide 1-inch-diameter foot, since roller cams aren't allowed in Grand National racing. Bud remains mum when it comes to specific cam grinds; although he does lean toward Crane, certain Competition Cams and Crower grinds have also been run successfully.

With large 2.19 intake/1.71 exhaust valve sizes, the use of lightweight titanium in valve construction is considered mandatory. TRW makes the intake and Del West the exhausts, with lash caps being required on the latter since it does not come with a hardened tip. Diamond Racing (part No. 3309) dual valve springs with dampner are used with Diamond's titanium retainer; with an installed height of 1.850 inches, they offer 150 pounds of seat pressure and 400 pounds "over the nose." Crane roller rockers and 5/16-inch-OD, stock-length Crane pushrods complete the valvetrain. The specific rocker arm ratio run (1.6:1, 1.65:1, 1.73:1) depends on the track. A Cloyes Tru-Roller chain helps keep the cam in time, and the heads are bolted to the block atop either an old Ford Muscle Parts head gasket (D3ZX-6051-AA) or McCord equivalent.

As is the case with the rest of this 500-mile engine, the valve job is designed as much for durability as it is for peak horsepower. A standard three-angle valve job is used, with 60-degree bottom, 45-degree seat, and 30-degree top angles. The intake seat width is usually around .070-inch, while the exhausts utilize a slightly wider .080-inch seat. A .030-inch top cut and full-blended radius seems to help the exhaust, but not the intakes. Perfect Circle seals are used on the intake valves only, since the hot exhaust side requires additional oil. New heads retain stock guides that are replaced after wear occurs by steel guides machined to offer .0022 to .0023-inch of clearance.

Considerable head porting was needed to make the old iron heads work. On the intake side, the bottom floor is filled up to a height of approximately ⅜-inch, which picks up flow below .400-inch lift without sacrificing any top end power. More radical surgery is performed on the exhaust side, where the exhaust port is cut away entirely and replaced by a fabricated plate with a recontoured port. The combustion chamber is also opened up near the spark plug, as

POWERED by FORD

shown in the photos.

Two different intake manifolds are used: an Edelbrock Scorpion with spacer plate on the ½ to 1-mile short tracks and Bud Moore's own two-piece box intake on the longer 1½ to 2½-mile super speedways. Both are matched to the reshaped intake port and further subjected to extensive (but undisclosed) internal rework. A highly modified Holley 750 double-pumper is used on the Scorpion, but the Moore intake accepts only the 4500 Dominator whose venturis are currently severely restricted by NASCAR rules.

Ignition is handled by Ford's old (but proven) electronic 429 racing system. Considerable research has gone into the exhaust system. Moore has found that changing the primary pipe diameter in midstream—say, starting with a 2-inch OD pipe at the flange and then transitioning it into a larger 2⅛-inch pipe before the collector is reached—can cut up to ½-second per lap on the super speedways. Shifting the point where the transition occurs also allows fine tuning of the torque curve. Naturally, research continues in this area.

Innovations such as these have allowed Bud Moore to remain competitive through the long, lean years. Now, thanks to the new Ford heavy-duty parts combined with sleek all-new sheetmetal, Bud ought to have the other guys seeing Ford blue. **HR**

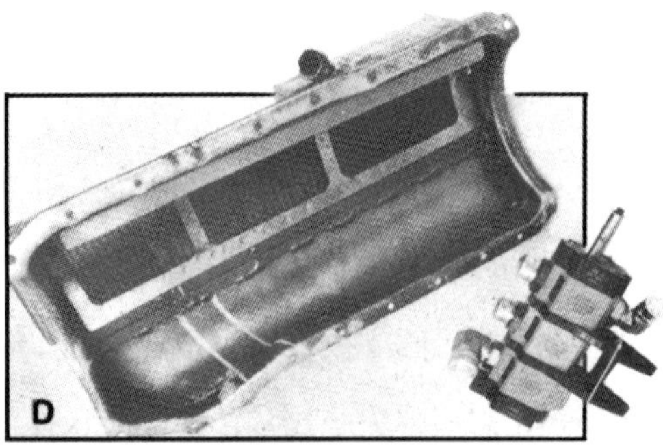

D *A Weaver dry sump oil system with three-stage pump keeps the engine alive while reducing ground clearance. The first stage pumps oil through the oil cooler into the engine, while the second and third stages scavenge the oil back to a trunk-mounted reservoir. Oil enters the engine via an adapter that hooks up to the stock oil filter mounting boss. The pan is made by Weaver to Moore's specs.*

E *Machining down the mushroom tappets' internal wafer washer **(arrow)** so the groove is only .006-inch deep is one way Moore restricts top end oiling.*

F *Quarter Master built this ultra-light triple-disc clutch setup that bolts to a standard automatic trans flex plate used in lieu of a conventional flywheel.*

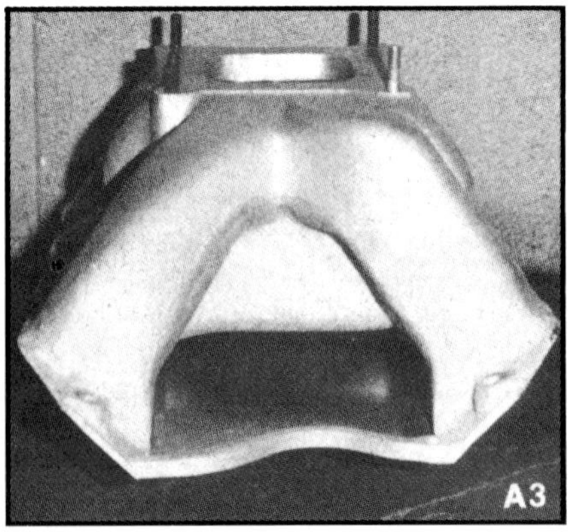

A *Depending on the track, either the Edelbrock Scorpion **(1)** with 2-inch spacer or Bud Moore's own "box-type" intake **(2)** is used, the latter offering more top-end for the super speedway while still holding up well in the mid-range. Ford's new prototype one-piece intake **(3)** is a descendant of the Moore piece, which in turn evolved from the old 302 Trans-Am mini-plenum factory intake. Since the plenum is reduced in size for improved throttle response, the intake-to-head bolts can now be reached externally, eliminating the need for the two-piece design.*

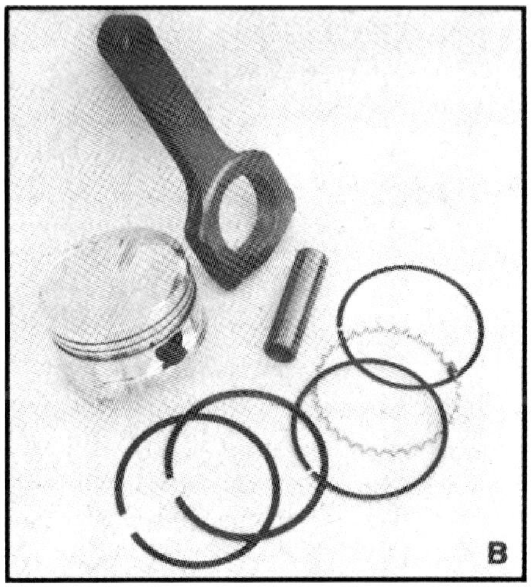

B *Custom Venolia pistons designed by Moore have a special dome to match the Moore-modified combustion chamber; they're used with Carillo rods, Precision Machine Shop pins, and Speed-Pro rings.*

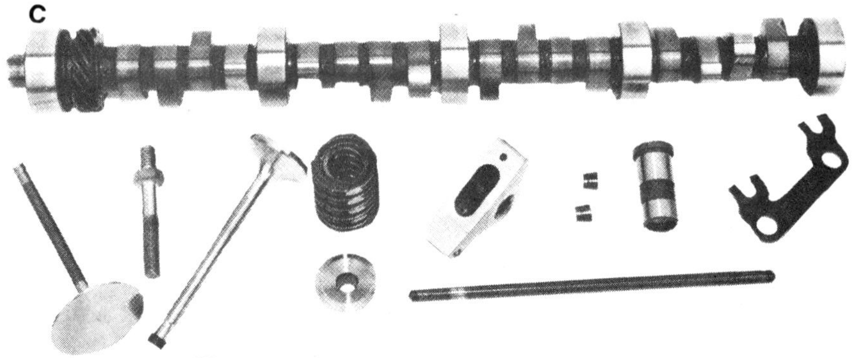

C *Since NASCAR still bans roller cams, Johnson mushroom tappets marketed by Speed-Pro are used to actuate Crane heat-treated pushrods, Crane roller rockers, Del West titanium exhaust valves, and TRW titanium intake valves. A lash cap is required with the non-hardened tip Del West valve. The Diamond valve spring and retainer is kept in place by stock Ford keepers. Manley supplies the stepped pushrod guide plates.*

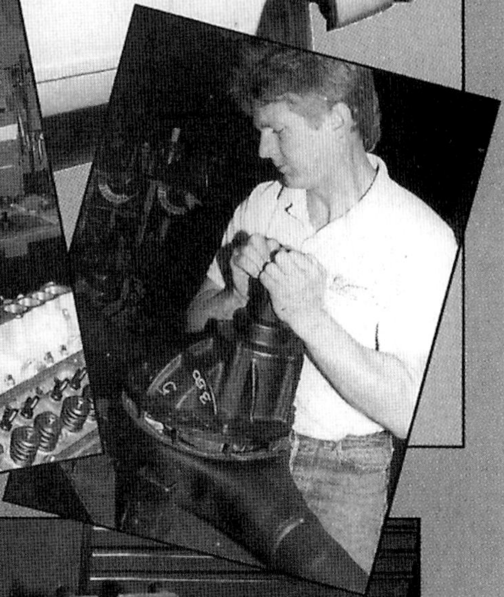

PROJECT BABY 'BOLT

DRIVE-LIGHTNING FOR SON OF THUNDERBOLT

By Joe Pettitt

Last month, we gave you a taste of this year's General Tire-sponsored Special Events grand-prize giveaway car—the Baby 'Bolt '64 Fairlane built by Scott Sullivan. This month, we're going to focus on the powertrain buildup.

The key player in this step of the project is Kaufmann Products, which modified a Ford 5-liter H.O. fuelie motor to generate approximately 300 hp at 5000 rpm and 310 lbs.-ft. of torque at 3500 rpm. The engine is built off a low-mileage short-block that was thoroughly cleaned and checked for mechanical soundness. The main and rod bearings were inspected and found to be perfect, as were the cylinder walls and pistons. Art Carr Performance Transmission Products also participated with a rebuilt C4 automatic to handle the power of the Kaufmann Stage III Ford fuelie. In addition, Currie Enterprises built a narrowed Ford 9-inch rearend to stand up to the hard "leaves" we anticipate the car will have to endure.

And because we tend to hammer the Pro Street crowd about being able to idle in traffic without overheat-

We begin with a thorough cleaning of the block deck and other critical sealing areas. The next step is to install the cam.

The Kaufmann Stage III package comes with a hydraulic roller cam that specs out at .500-inch lift, with advertised 270 intake and 280 exhaust duration and a lobe-separation angle of 114 degrees to generate good idle quality. At .050-inch tappet lift, the cam has 212 degrees of intake duration and 222 degrees of exhaust duration. Kaufmann Products also included a double roller chain and gear to ensure that the two-degree-advanced installation on the cam timing remains accurate.

Most engines are limited by the amount of air the heads flow. The key, then, to making good power is getting the heads to generate serious airflow and yet maintain enough velocity to produce good torque. The heads are original '66 4V 289 iron Ford small-block castings that have received extensive massaging. The combustion chamber has been recontoured and polished to promote an even flame front and high-swirl characteristics. Kaufmann Products swirl-polished stainless valves (1.900-inch intake, 1.600-inch exhaust) are designed to enhance flow, and a three-angle valve job is "standard equipment."

The intake port is matched to the gasket and intake runner. You can see that material was removed to about one inch into the port. This opens the port slightly, but more important, the flow of air is undisturbed by gasket overhang or intake manifold runner to port wall misalignment. With the Kaufmann valves at .500-lift, the intake flows 92 cfm.

On the exhaust side, extensive bowlwork is performed to improve the flow, which is essential because of the small production Ford port. If done correctly (and the guys at Kaufmann Products know Fords), you can generate some impressive flow numbers. Getting this medium-size port to flow 64 cfm is no easy task, but they do it.

The rest of the buildup involved dressing the engine with accessories and buttoning it up. A stock computer will control the engine with a Kaufmann Products loom that eliminates the emissions-related controls. The computer still monitors the oxygen sensors, manifold-absolute-pressure (MAP) sensor, engine coolant, air-charge temperature, and throttle-position sensors. With 10:1 compression, good flowing heads, and fuel injection, this motor should be crisp and powerful.

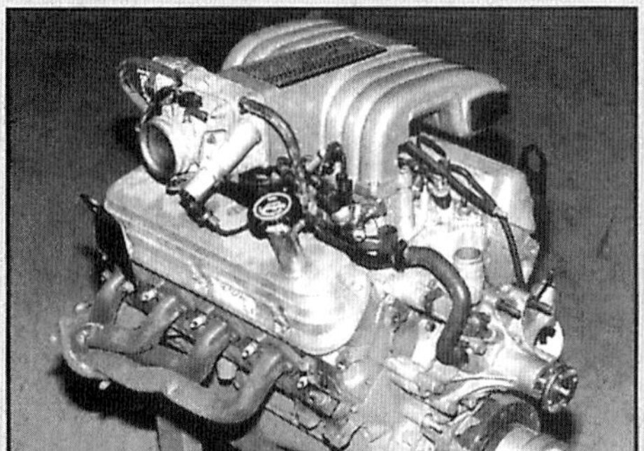

After installing the stock roller lifters and retaining device, new studs in the heads, and another coat of silcone on the intake-manifold side of the intake gasket, Shannon Laird installs the lower portion of the intake. Both the upper and lower portions of the intake track are massaged to increase the flow.

Art Carr Performance Transmission Products was selected to build the transmission for Baby 'Bolt. Art Carr is a savvy racer and in fact helped build one of the original '64 Thunderbolt Fairlanes. One of his street/strip transmissions backs the Kaufmann motor to provide snappy shifts and full driver control over shift points with heavy-duty clutches and bands.

Art Carr uses heavy-duty clutch plates. They are made from the highest-grade friction material and feature a cross-hatch groove for smooth, even engagement, better grip, and durability.

ing, Griffin Radiator Manufacturing built a trick aluminum radiator for Baby 'Bolt. The company claims its two-row, 1¼-inch tube radiator cools as well as a seven- to eight-row copper/brass unit. For a Pro Street car with a big-inch engine, the extra clearance is a definite advantage. But since the Griffin radiator is only two rows thick, it offers less resistance to airflow, so

In order to build transmissions that will stand up to high-torque motors, Art Carr uses high-grade friction materials for the intermediate bands. They are much stronger than the flexible stock unit shown here.

Each Art Carr transmission is blueprinted and built with the highest-quality parts to increase its torque-handling capabilities. After assembly, the transmissions are dyno tested to make sure the shift points are correct and everything functions properly.

Torque converters are an essential part of a well-tuned automatic, so Art Carr included an 11-inch converter to generate a stall speed of around 2200 rpm. Its Super Torque street converter is a product of off-road-racing research that works perfectly on the street. The precise angle of the vanes and stator give exceptional torque multiplication and improved lock-up at high rpm. Notice that the welds on the converter blades are furnace-brazed, which reduces turbulence and friction within the transmission fluid. Heavy-duty Torrington-type needle bearings are used to handle high-thrust loads.

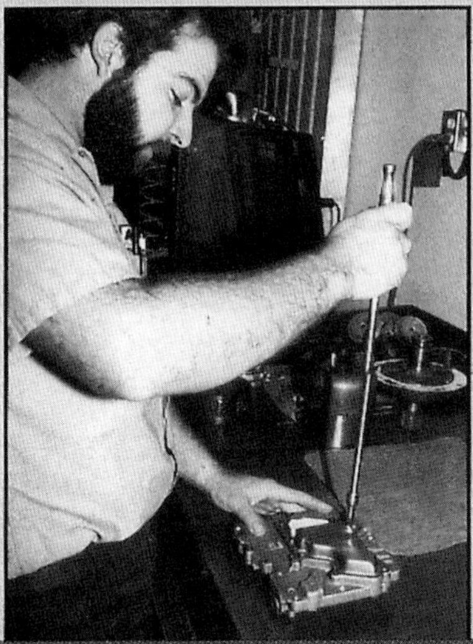

Greg Antonucci assembles the valve body with the Art Carr Hi-Tech kit. The kit reprograms the shift-control mechanism, yielding firm gear changes and quicker manual control of gear changes, when desired.

An added feature of Art Carr converters is a burnishing process on the hub. Burnishing hardens and polishes the hub, making it stronger and also increasing seal life because the polished surface generates less friction. In addition, all Art Carr street converters are electronically balanced to within five grams.

Currie builds everything in-house, so it can control the quality. The splines are machined in, a process that provides greater strength than other methods.

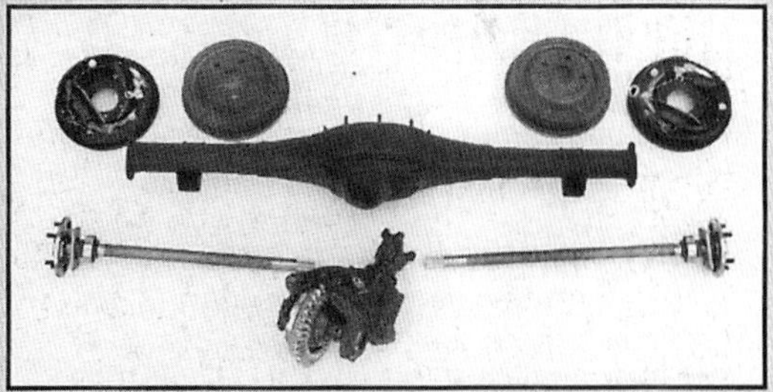

Currie Enterprises was enlisted to build a narrowed Ford 9-inch rearend that would stand up to the hot Kaufmann small-block. The stout rearend has a 3.50 gear and 31-spline, high-grade steel axles all wrapped up in a swedged and TIG-welded axle housing. The Currie-supplied drum brakes feature heavy-duty semimetallic brake linings for positive stops from high-speed quarter-mile blasts.

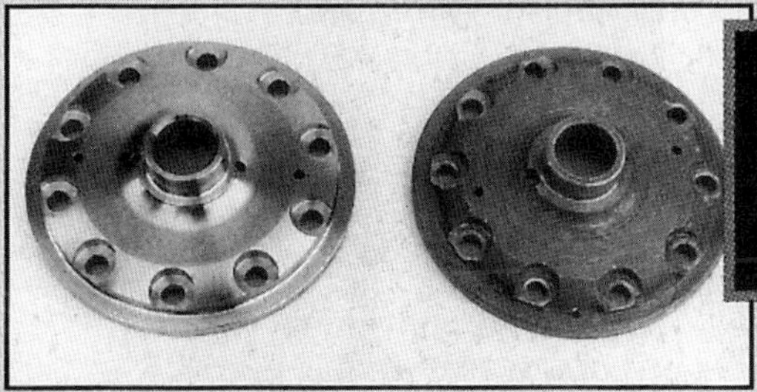

An alloy-steel side plate is part of the package. It resists flexing under high loads and keeps the gear teeth aligned properly for more efficient transfer of power and longevity in severe-duty conditions.

it tends to flow more air in slow-speed situations (such as idling through the fairgrounds), which will help keep the engine at a reasonable operating temperature in prolonged slow-speed environments.

With all of the major components in place, let's hotfoot it over to Kaufmann Products in Downey, California, and follow along as Chris Kaufmann supervises the building of a bit of Blue Oval thunder for Baby 'Bolt. **HR**

SOURCES

Art Carr Performance Transmission Products
Dept. HR
10575 Bechler River Ave.
Fountain Valley, CA 92708
(714) 962-6655

Currie Enterprises
Dept. HR
1480-B N. Tustin Ave.
Anaheim, CA 92807
(714) 528-6957

Griffin Radiator Manufacturing, Inc.
Dept. HR
Rt. 1, Box 66
Townville, SC 29689
(800) 722-3723
(800) RACE-RAD

Kaufmann Products, Inc.
Dept. HR
12400 Benedict
Downey, CA 90242
(213) 803-5531

KING OF

By Marlan Davis

25 Years of MUSTANGS

STROKING THE 351W FORD TO 372 CUBIC INCHES

At the cost of $300 to $400 over the amount normally spent on a small-block Ford overhaul, you can stroke a 351W engine .140 inch and wind up with 372 cubic inches (including the typical .030-inch overbore). No exotic parts are required—just an off-the-shelf 351W crank, Chrysler 318 connecting rods, and forged small-block Chevy "turbo" pistons. According to Total Performance owner John Vermeersch—Ford performance specialist and one of the originators of this combination—the net result is an engine that is 25 to 30 percent stronger than standard 351W-based motors in both horsepower and torque over a much broader rpm range. Depending on the rest of the parts combination (cam, cylinder heads, compression, induction, and exhaust), you'll end up with either a street-king combo that fits within the standard small-block Ford engine envelope and accepts normal 302/351W accessories, or a competitive budget entry for the local drag or circle track wars. Sound intriguing? Let's get down to the long strokes.

A 302, 351W... or 372? Because it accepts standard Windsor accessories and fits in the same envelope, only the engine builder knows for sure!

BOTTOM END

John begins with a standard 3.50-inch-stroke, 351W cast-iron crank, offset-grinding its stock 2.311-inch rod journals down to the Chrysler 318 rod's 2.1245-inch size. In theory, a potential .186-inch stroke increase is obtainable via the change in journal size, but in this case the actual offset grind is limited to produce only a .140-inch increase, thereby allowing utilization of off-the-shelf pistons while also leaving .046-inch material leeway for cleanup of any damaged journals. There are several different Chrysler small-block rods; so far, John has verified that only 318 forging No. 3418645 fits in the Ford stroker motor without extensive rework, although other, similar Mopar "LA" V8 non-offset rods should also work. Retain the rod on the crank with high-quality aftermarket rod bolts and use premium tri-metal rod bearings designed for 318 Chrysler applications.

At the other end of the Chrysler rod, when using standard Wedge (non-Cleveland style) small-block Ford heads, a TRW L2441F-30 small-block Chevy "turbo" piston (or equivalent) puts the deck height in about the right place on the Windsor block, yielding approximately 9:1 compression (chart B), depending on the particular head casting chosen. This piston must be milled for use in '69-'70 Windsor blocks, which had a lower deck height than later castings. In any case, the piston dish edge adjacent to the quench area must be laid back about ⅛ inch for valve clearance; the exact amount depends on the camshaft and head gasket, requiring that this dimension be checked by "claying" the piston during trial assembly.

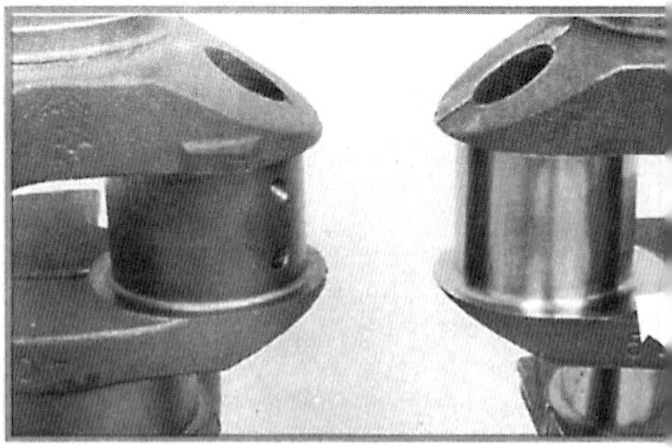

Offset-grinding a 351W crank's rod journals down to the Chrysler 318's dimensions increases the stroke .140 inch (right) *and permits the use of slightly modified 318 rods. Because no material is removed from the counterweight, heavy "Mallory" metal isn't required when balancing the engine. To preserve adequate crank structural integrity with the new downsized rod journal, many crank grinders machine a generous fillet radius at the rod journal-to-crank throw (counterweight) interface* (arrow).

THE WINDSORS

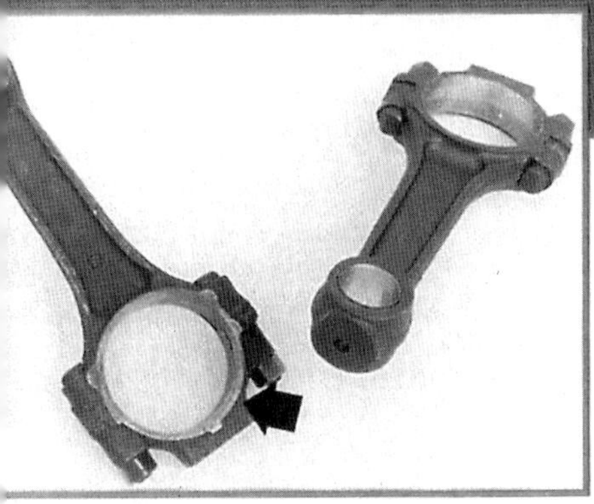

The large crank journal fillet radius in turn requires the recommended Chrysler 318 rods' big ends (arrow) and rod bearing edges to be correspondingly chamferred for clearance. The small ends on these rods are bushed to reduce their diameter down to that of the Ford or Chevy piston pins. These floating pins also require drilling a hole through the rod top for pin oiling.

The rods' small ends are narrowed approximately .060 inch per side to fit between the Chevy or Ford piston bosses. Removing material from the pistons' inner pin bosses is seemingly an alternative approach, but that excessively weakens the aluminum piston's pin boss.

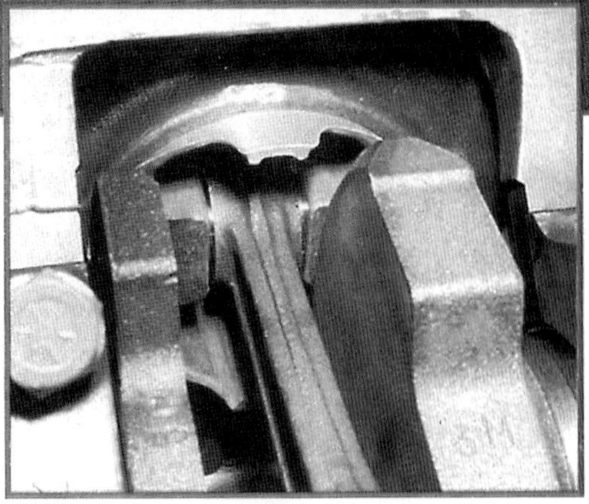

While not usually a problem with this combo, on a stroker engine it is always advisable to mock up the assembly in the block and verify that the nonstock piston/rod combo clears the crank throw, oil pan rails, bottom of the cylinder bore, and camshaft.

Because the Chevy piston uses .927-inch-diameter pins, while the Chrysler rod's small end is machined for a .9842-inch piston pin, a bronze bushing must be installed in the rod to make up the difference—but that's required anyway when using floating piston pins such as those furnished with the L2441 piston. The Chrysler rod's small end is also too wide in relation to the piston pin bosses' inner edges. It must be narrowed approximately .060 inch per side (.120 inch total).

Cleveland-style heads require pistons with corresponding canted valve reliefs. In this stroker motor, pistons originally designed for Boss 302 engines produce the correct deck height. Their .912-inch-diameter pins once again require bushing the Chrysler rods' small ends. Also as above, assembly clearances must be checked and any appropriate machining performed.

A

372 FORD POWER COMBOS

	STREET	STREET/STRIP		DRAG		OVAL TRACK
ESTIMATED HP	335-350	380-400		440-460		450-475
POWER RANGE (rpm)	3500-5500	4000-6200		4500-6800		4500-6800
CARBURETOR						
Size	750 cfm	750 cfm		800 cfm		800 cfm
Type	4-barrel w/vacuum secondaries	4-barrel w/mechanical secondaries		4-barrel w/mechanical secondaries		4-barrel w/mechanical secondaries
Brand	Holley 0-83310 (auto trans) Holley 0-3310 (manual trans)	Holley 0-4779		Holley 0-4780		Holley 0-4780
INTAKE MANIFOLD						
Type	Dual-plane	Single-plane		Single-plane		Single-plane
Brand	Edelbrock Performer 351W Part No. 2181 (non-EGR) Part No. 3781 (EGR)	"Wedge"-head: Edelbrock Torker II 351W (part No. 5081)	"2V" Cleveland head: B&A Track Boss 351W (part No. 201)	"2V" Cleveland head: B&A Track Boss 351W (part No. 201)	Alloy high-port Cleveland head: Motorsport M-9424-A351	Short-track: Motorsport M-9424-A351 Long-track: Edelbrock Victor 351-GN Part No. 2988 or Motorsport M-9424-C351
CAMSHAFT						
Type	Hydraulic	Hydraulic		Mechanical		Mechanical
Duration (advertised)	280-290 degrees	290-305 degrees		295-306 degrees		295-310 degrees
Duration (.050 tappet lift)	225-235 degrees	235-250 degrees		255-265 degrees		260-270 degrees
Valve lift	.490-.520 inch	.535-.560 inch		.590-.640 inch		.640-.648 inch
HEADERS						
Type	4 into 1 or tri-Y	4 into 1		4 into 1		4 into 1
Primary tube diameter	1½-1⅝ inches	1⅝-1¾ inches		1⅞-2 inches		1⅞-2 inches
Collector diameter	3 inches	3 to 3½ inches		3½ inches		3½ inches
PISTONS						
Type	Forged, dished	"Wedge"-head: Forged, dished	"2V" Cleveland head: Forged, domed, canted valve reliefs	Forged, domed, canted valve reliefs		Forged, domed, canted valve reliefs
Brand	TRW L2441F-30 or Speed-Pro 7058PA-30 **(A)**	TRW L2441F-30, or Speed-Pro 7048PA-30 **(A)**	TRW L2324F-30 **(B)**, Speed-Pro 7036P-30 **(C)**	TRW L2324F-30 **(B)** Speed-Pro 7036P-30 **(C)**		TRW L2324F-30 **(B)** Speed-Pro 7036P-30 **(C)**
Compression ratio	(For compression ratio data, see chart B, "Compression Ratio Combinations." Letters following pistons above, and heads below, correspond to type letters in chart B.)					
HEADS						
Type	Windsor: '69-'71 iron or Motorsport M-6049-J302 alloy **(Type W/C)** Windsor: '77-up iron **(Type W/O)**	Motorsport alloy M-6049-J302 **(Type W/C)**	Cleveland 2V **(Type C/2V)**	Cleveland 2V **(Type C/2V)**	Motorsport alloy high-port M-6049-C302B or equivalent **(Type C/M)**	Motorsport alloy high-port M-6049-C302B (or earlier alloy equivalent) **(Type C/M)**
Intake valve diameter	1.900 (oversize for iron) (part No. M-6507-B321) 1.940 (std. on M-6049-J302) (part No. M-6507-A302)	1.94 (with M-6049-J302) (part No. M-6507-A302) 2.05 (with Cleveland 2V head) (production-type)		2.05 (Cleveland 2V head)	2.15 or 2.19 (Motorsport high-port)	2.15 or 2.19 (Motorsport alloy high-port head) (special aftermarket race type)
Exhaust valve diameter	1.625 (oversize for iron) (part No. M-6505-A321) 1.600 (std. on M-6049-J302) (part No. M-6507-A302)	1.600 (with M-6049-J302) (part No. M-6507-A302) 2.05 (with Cleveland 2V head) (production type)		2.05 (Cleveland 2V head)	2.15 or 2.19 (Motorsport high-port	2.15 or 2.19 (Motorsport alloy high-port head) (special aftermarket race type)
IGNITION						
Type	Breakerless, magnetic impulse, with CD multispark amplification					
Brand	Ford Duraspark distributor with MSD-6T trigger					

TOP END

There are many cylinder head variations available for use on Ford small-blocks. Trying to describe all the variations would entail writing a book! Here we have space to look at only some of the more common cylinder head castings.

For the typical street package, the most commonly available 351W heads are the 68cc "open-chamber" designs used from '77 up. Most of these have small 1.78-inch-intake/1.45-inch-exhaust valves. When upgraded to the earlier 351W, 1.846-inch intake/1.546-inch exhaust; "GT-40"-style, 1.90-inch intake/1.625-inch exhaust; or even small-block, Chevy-style, 1.94-inch intake/1.60-inch exhaust—and then treated to a good three-angle valve job, pocket porting, exhaust-side bowl work, air-injection bump removal, and intake manifold port-match—they make a fine street head, albeit at some sacrifice in compression ratio. If you can find a set, the earlier '69-'74 351W heads are a better choice, as they have slightly larger ports and valves, plus smaller 58 or 60cc chambers.

B COMPRESSION RATIO COMBINATIONS

Except as noted, all dimensions in inches or fractions thereof.

CYLINDER HEAD		HEAD GASKET			PISTON				COMPRESSION	
TYPE	CHAMBER VOL. (cc)	FORD MOTOR-SPORT PART NO.	COMPRESSED THICKNESS	VOL. (cc)	TYPE	COMPRESSION DISTANCE (design)	DOME (+) OR DISH (-) VOL. (cc)	DECK HEIGHT (below)	RATIO	NOTES
W/C	58	M-6051-A302	.042	9.1	A	1.560	-21.1	0	9.2-9.7:1	1
W/O	68	M-6051-A302	.042	9.1	A	1.560	-21.1	0	8.75-9.15:1	1
C/2V	76.2	M-6051-A331	.040	8.8	B	1.530	+8.9	.007	10.81:1	2
					C	1.530	+12.1	.007	11.23:1	2
C/M	65	M-6051-A331	.040	8.8	B	1.530	+8.9	.007	12.47:1	2
					C	1.530	+12.1	.007	13.05:1	2

NOTES

1. Actual CR varies according to amount of piston deck and/or valve notch machining required. Listed "0" deck is based on '71-up 351W block, 9.503 inches nominal design crankshaft centerline to deck. Listed *piston* deck must be machined .023 inch to achieve "0" deck with '69-'70 block, 9.480 inches nominal design crankshaft centerline to deck; in that case, approx. CR is 8.75:1. If piston is used in '69-'70 block, but machined only .016 inch, to achieve minimum acceptable .035-inch piston deck-to-head deck clearance, nominal compression is slightly over 9:1. These are nominal values only; always mock-up assembly and verify actual ratio.

2. Compression ratio and deck height based on '69-'70 block. Later *blocks* require decking .023 inch to achieve these values.

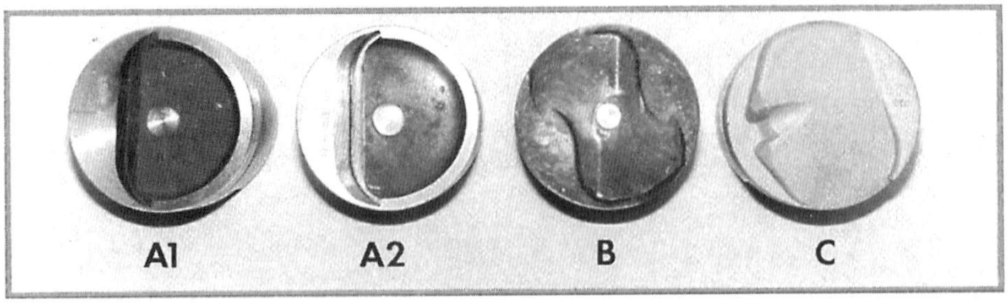

A "Chevy turbo" piston (TRW L2441F or Speed-Pro 7058PA) (A1) *produces street-able compression ratios with the Chrysler rod when using Windsor heads, but the quench area must be laid back approximately ⅛ inch* (A2) *for valve clearance. For competition use with Cleveland-style heads, use low- or high-dome* (B and C, respectively) *Boss 302-style pistons.*

Street/strip engines should use a Ford Motorsport aluminum, Windsor-style head M-6049-J302, which is worth 30 hp over typical late-model iron heads. Another alternative for a hotter motor are old Cleveland "2V" castings. The small chambers common to most "4V" Cleveland heads raise compression too high on a street engine, plus their huge valves and ports seriously hurt the low end and midrange. However, such 4V heads, or even a set of the Motorsport high-port Cleveland/SVO aluminum castings, are a good choice for drag- or circle-track-only engines. Machining is required to install Cleveland heads on Windsor blocks. Basically, the existing deck outlet water passage must be plugged, and a new passage drilled in the head intake face. Also required are special 2V or 4V intake manifolds, plus (in the case of 4V or Motorsport heads) a cast-aluminum water crossover manifold M-8592-A351. For head machining instructions, see (for iron or alloy heads) Ford/Motorsport *1989 Performance Equipment Catalog* (M-1989-A1) or (iron heads only) HOT ROD Magazine's Pit Stop, "Getting a Head," January '89, pp. 136 and 138.

Naturally, any of these heads must be used with complementary camshaft and valvetrain components, which vary according to head type and intended application. Generally, for street use with an automatic trans and/or power accessories, a 280-degree advertised-duration hydraulic cam proves to be the best choice. Manual trans cars can move up to the 290-degree level. In the absence of a dyno to test different combos, generally the cam with the most lift in relation to the desired advertised duration offers the best performance. Typical camshaft specs and recommended rocker arm/pushrod combos needed with various heads are shown in charts A and C, respectively.

Head choices include iron Windsor junkyard castings, new Motorsport aluminum Windsor, open-chamber Cleveland 2V, and (for competition use only) closed chamber/big port-and-valve Cleveland 4V.

Most Ford small-block heads lack provisions for the adjustable valvetrain that's often required to compensate for head swaps and/or high-lift camshafts. The '69-'76 351W heads were machined for a press-in, positive-stop rocker arm stud, and they used "rail" rockers that maintained proper pushrod alignment in lieu of guideplates and/or close-tolerance holes machined in the head. This system will work with moderate-performance aftermarket hydraulic cams. With a hydraulic cam using antipump-up lifters and/or having more than .530- to .540-inch lift, or any mechanical cam, an adjustable valvetrain is required—the press-in studs must be removed, and the stud bosses milled and tapped for screw-in studs that work with guideplates, hardened pushrods, and conventional (nonrail) rocker arms or roller rocker arms.

Paired "pedestal" rockers retained by bolts and a fulcrum guide are used on '77-up 351W heads. Ford Motorsport offers bolt-on roller rockers that replace the stock stamped rockers. However, neither the stock pedestal nor the "bolt-in" replacement rollers allow for lash adjustment. By contrast, Crane offers a bolt-on stud kit with complementary roller rockers that allows conversion of this type of head to an adjustable valvetrain, with no machining required. Because the replacement rocker stud's in-head diameter is only 5⁄16-inch (so it'll screw into the existing mounting hole), the kit is not recommended for use with mechanical cams or extreme high-rpm use. With the latter, machine the heads for full-diameter, conventional 7⁄16-inch rocker arm studs and guideplates (milling data, chart D).

Non-Boss Cleveland-style heads have no means of valvetrain adjustment, either. The same sort of conversion pieces are available from the companies listed above for these heads as

Early closed-chamber Windsor heads (left) *have smaller combustion chambers and larger ports and valves than the later '77-up open-chamber design* (right), *but all the later castings feature unleaded-gas-compatible, induction-hardened seats for stock-size valves.*

Minor water-passage machine work is needed to install Cleveland heads on Windsor blocks. The huge port and valve, closed-chamber 4V/Boss/Motorsport type (left) *are reserved for professional competition engines, but the more reasonable open-chamber 2V style* (right) *makes a fine choice on weekend warrior rigs.*

well. Aluminum Motorsport heads (both Wedge and Cleveland styles) come machined for adjustable valvetrains.

Just as the camshaft, valvetrain, and cylinder head must complement each other, the intake manifold and carb must also be matched for best results. On the street, a dual plane such as the Edelbrock Performer works well. Dual-purpose or full-race cars need a single-plane intake—the specific type varying with cylinder head choice and application (chart A). Holley's 0-3310 750-cfm vacuum secondary carb works well on the street. Part No. 0-83310 has provisions to accept Ford automatic transmission kick-down levers. More radical cars use Holley double-pumper series carburetors.

Once the mixture is in the chamber, it should be ignited by a good high-capacity ignition system. If you like Fred Flintstone, then the old points distributor is right up your alley. Those who see the light will opt for a Ford Duraspark distributor controlled by an MSD-6T, with high-quality, helically wound suppression wires. Exhaust the spent mixture using $1\frac{1}{2}$- to $1\frac{5}{8}$-inch headers on the street, $1\frac{3}{4}$ inches for street/strip cars, and $1\frac{7}{8}$ to two inches for full-competition applications. Arrive at final collector length by bolting on an extra collector extension with a longitudinal white paint stripe on it. Make several runs or laps; inspect the tube, and cut it off at the point where the paint hasn't burned off. Due to the Windsor block's unique deck height, headers designed for Cleveland heads may require modification to fit into otherwise stock chassis.

With a relatively mild cam and iron Windsor heads, this combo should produce in the neighborhood of 340 hp, with a powerband in the 3500- to 5500-rpm range. At the other end of the spectrum, a high-compression competition version with Cleveland-style heads will easily put out 450 hp. Beyond that level, the stock iron crank with its large 3.00-inch-o.d. main journals becomes the limiting factor, and you should consider moving up to the thick-wall SVO block that's machined for the smaller main-bearing Cleveland crank and achieving the desired displacement by boring and stroking. But for the stoplight grand prix, Saturday night brackets, or short-track "claimer" race, these packages provide decent power and torque for a very reasonable price. So for those of you who'd rather have Windsor blue blood than join the "General's" army, there's finally an alternative to the "other guy's" 383 small-block. **HR**

C 351/372W PERFORMANCE ROCKER ARM AND PUSHROD COMBOS

All dimensions in inches or fractions thereof.
All part numbers are Ford/Motorsport unless otherwise indicated.
Because of differing cam base circles, block and head surfacing dimensions, head gasket variations, and valve stem length and seat locations, it is strongly recommended that all high-perf engines use an adjustable valvetrain.

CYLINDER HEAD	ROCKER ARM			ROCKER ARM STUD	GUIDEPLATE	PUSHROD	
	Type	Adjustable?	Part No.			Length	Part No.
Pre-'77 iron Windsor	Non-rail, cast iron	Yes	M-6564-A321	M-6527-B311 (Note 1)	M-6566-A311	8.152	M-6565-A332
	Roller		*M-6564-A351				
'77-up iron Windsor	Improved-strength stamped steel	No	M-6564-A302	(Bolt and fulcrum included w/ rocker arm)	Not used	8.182	M-6565-B332
	Roller		*M-6564-B351				
	Non-rail, cast iron	Yes (Note 2)	M-6564-A321	Crane bolt-on conversion kit 36655-16	Not used		
	Roller		*Crane 36750-16				
	Roller	Yes	*M-6564-A351	M-6527-B311 (Note 3)	M-6566-A311	8.152	M-6565-A332
	Non-rail, cast iron		M-6564-A321				
Aluminum Windsor M-6049-J302	Non-rail, cast iron	Yes	M-6564-A321	M-6527-B311	M-6566-A311	7.601	M-6565-B322
	Roller		*M-6564-A351				
Cleveland, except Boss or SVO aluminum high-port	Roller	No	*M-6564-C351	(Bolt & pedestal included w/kit)	Not used	8.800	M-6565-D342
	Roller	Yes (Note 2)	*Crane 27750-16	Crane bolt-on conversion kit 52655-16	Included w/Crane kit		
	Stamped steel		M-6564-A322				
All Cleveland iron and aluminum	Stamped steel	Yes	M-6564-A322	M-6527-A311 (Note 4)	M-6566-A302		
	Roller		*M-6564-A460				

*Complete set of 16. Other rockers sold individually.
1. Must remove pressed-in bottle-neck stud, and tap stud hole for 7/16-14 thread.
2. Not recommended with valvespring pressure over 320 pounds.
3. Must machine cylinder head according to Chart D.
4. Must machine iron (except Boss) heads according to Chart D.

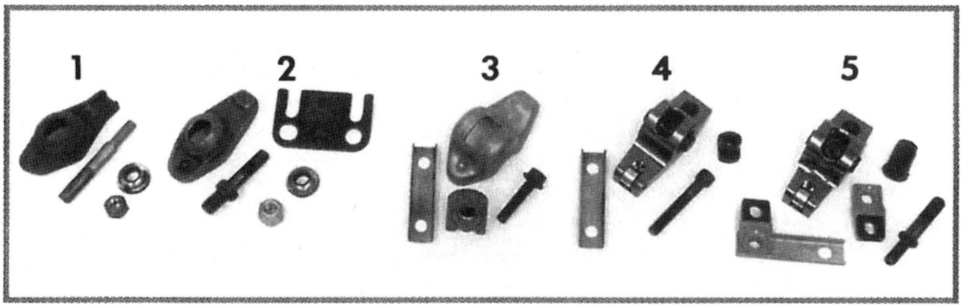

Several different rocker arm setups have been used on Windsor cylinder heads: '69-'76 used nonadjustable rail rockers with a positive-stop bottle-neck stud (1)*; to convert to adjustable valvetrain nonrail (or aftermarket roller) rockers, screw-in studs, hardened pushrods, and guideplates are required* (2)*. The '77-up used nonadjustable paired pedestal rockers, fulcrum guides, and bolts* (3)*. Ford Motorsport offers a nonadjustable bolt-on roller rocker kit* (4)*, while Crane and Comp Cams have an* ***adjustable*** *bolt-on roller rocker kit* (5) *that's suitable for use with valvesprings under 320 pounds.*

Suitable intake manifolds include (from left) *Edelbrock Performer (street use, Windsor heads), Edelbrock Torker II (street/strip use, Windsor heads), and B&A Ford Track Boss 351W (strip use, 2V Cleveland heads).*

D HEAD MODIFICATION FOR MECHANICAL CAM

To modify nonadjustable rocker arm heads to work with a mechanical cam: Machine at right angles to the existing hole (not the bottom of the head), then drill and tap hole for 7/16-14 thread.

CYLINDER HEAD	MILLING DIMENSION	MEASURE FROM
Windsor	.230 inch	'69-'76: Top of stud boss '77-up: Top of pedestal
Cleveland	.300 inch	Bottom of pedestal slot

SOURCES

B&A Ford Performance
Dept. HR
114 Towson Ave.
Fort Smith, AR 72901
(501) 782-6979

Ford Motor Co.—SVO
Dept. HR
17000 Southfield Rd.
Allen Park, MI 48101
(313) 337-1356

Total Performance
Dept. HR
44020 N. Groesbeck Hwy.
Mount Clemens, MI 48043
(313) 468-FORD

Mustang Magic BLUEPRINTI

By Bruce Caldwell

The 289 Ford is a great little engine. It can easily give 100,000 miles of relatively trouble-free service before needing an overhaul. The engine in our '65 Mustang fastback project car was just such a high-mileage 289, and although it was way out of tune, it still ran okay. We wanted more than just an okay engine; we wanted a strong street and occasional strip motor that would be up to the level of the rest of our Mustang revitalization.

Since the engine in our Mustang still ran well enough to get the car to and from the various shops it frequented during its rebuilding process, we decided to leave it alone and build another one. This method is especially good if the car you are rebuilding is your only source of transportation. Used 289s are inexpensive, so you should be able to pick up a core engine for under $200 in running condition and even less if it doesn't run. Our core engine was free (all we had to do was pull it out of a wrecked Mustang sitting in a friend's backyard), but it didn't run. Our bargain engine's demise was thought to be due to lack of oil pressure, but judging by the shape of the car it came out of, death could have been caused by a terminal case of sludge.

Jim McFarland of Edelbrock generously offered the services of Edelbrock's engine dyno, but when "the engine that ate Cleveland" showed up at their sanitary El Segundo, California, facility, everyone shielded their eyes and left the room. Luckily, noble development engineer Curt Hooker stepped forward and volunteered to work on it.

Disassembly of an engine is pretty straightforward: simply remove everything until you are down to the bare block. Along the way, inspect all the parts for wear and damage, mark and keep track of where every part goes for reassembly, and try not to damage anything through use of excessive force. Use a lot of coffee cans, boxes, or bags to segregate parts. Mark the containers, or draw diagrams of how things came apart. At the time you

tear down the engine, everything is fresh in your mind, but later you might forget where that strange part goes. If at all possible, steam-clean the engine before disassembly.

The basic sequence of disassembling a 289 Ford is just that—basic. Mount the engine on an engine stand for working ease; drain the oil; remove front-of-the-engine accessories like air conditioning, alternator, and power steering pump;

THE 289

TRANSFORMING A JUNKER INTO A SUPER STREETER

(1) This once-noble Mustang coupe gave its motor for our Project Mustang. The engine was free for the taking, and by rebuilding this engine, we were still able to drive our Mustang.

(2) Our free engine as delivered to Edelbrock's sparkling dyno facility. Not a pretty engine, but not unlike what most people start with, either.

(3) The engine was definitely not a cherry, as witnessed by the water pump leak that was repaired with body filler.

(4) Upon disassembly, we discovered many things about our motor, including the reason for its death—a broken oil pump shaft (arrow).

remove the exhaust manifolds and carburetor; with the valve covers out of the way, loosen and remove the rocker arms and the pushrods; and take off the cylinder heads. That takes care of the top of the engine.

Now, remove the oil pan and turn the engine over on the engine stand. Next, remove the oil pump. (The oil pump shaft was broken on our engine, which explained the loss of oil pressure.) Continue disassembly by removing the rod caps and bearings; removing the pistons; using a puller, removing the harmonic balancer from the front of the engine, which will make the timing chain and cover accessible; removing the main bearing caps and lifting out the crankshaft; and removing the camshaft and lifters. What you have left is a bare block.

Besides a bad oil pump, Curt's inspection revealed that our freebie engine also had bad bearings, clogged oil rings, a loose timing chain, and a crankshaft that was slightly out of round. The engine appeared basically sound and fine for rebuilding. It was a typical 100,000-mile-plus, used engine: tired out, gummed up, but plenty good for another 100,000 miles after a rebuild.

Next on the agenda was the all-important visit to the machine shop. We took our project Mustang engine to Valley Head Service's huge machine shop and engine-building facility at 19340 Londelius, Northridge, CA 91324, (213) 993-7000. Valley Head Service is equipped to handle everything from cleaning and machining to engine assembly and dyno-testing. Besides doing a tremendous amount of local business, Valley Head Service also provides mail-order service.

Valley Head Service's owner, Larry Ofria, explained what needed to be done to our engine. Rather than use a lot of exotic speed equipment, we decided to build a strong, reliable, balanced-and-blueprinted engine. This type of engine works beautifully with common aftermarket speed equipment, and if we later get the urge for more horsepower, the basic engine will be able to handle the extra load. Too many people build an engine backward; they add all the trick eyewash items but ignore the basic powerplant. All the camshaft and carburetion in the world won't do you any good if the bottom falls out of the motor.

The goals for our engine were as follows: power in the 275 to 300-horsepower range, plenty of bottom-end torque, quarter-mile

(5) Curt Hooker of Edelbrock checked the crank with a micrometer to see if the crankshaft needed to be reground. The crank was slightly off, so a regrind was ordered.

(6) All the machine work was done at Valley Head Service. Before the engine could be cleaned, all plugs had to be removed. Here the cam bearings are being driven out of the block.

BLUE-PRINTING THE 289

(7) Our slimey block was subjected to a long stay in the hot tank. A hot tank is like an oven which cleans with heat and a caustic solvent.

times in the high 13s to low 14s, 15 to 20-mile-per-gallon economy under everyday conditions on standard pump gasoline, smooth around-town operation, and good emissions levels. Both Jim and Larry felt the 289 Ford was easily capable of the above parameters.

Very few people can do their own machine work, so we'll just present an overview of what is involved in blueprinting an engine. The basic premise of blueprinting is to build a perfectly matched and balanced engine. Tolerances are closely adhered to, and cleanliness is paramount. Every part is inspected for weaknesses, and only perfect parts are used.

Before any machine work could be performed, our crusty engine needed a thorough cleaning. All freeze plugs and cam bearings had to be removed because they will dissolve during the hot-tanking process. All the parts were placed in the hot tank, which cleans by means of a heated and very caustic solvent. After the engine was through in the hot tank, it was rinsed, followed by additional cleaning of small passages with solvent and small bristle brushes. After everything was clean, it was "Magnafluxed." Magnafluxing is the cheapest engine insurance you can get, for you find weak parts before they are reinstalled in your engine, not when you make your first high-speed pass.

The main bearing bores were checked to see if it would be necessary to align-bore the engine. Larry said small-block Fords very seldom need align-boring (the odds are about 20 to 1). Our engine was within acceptable tolerances, so there was no need to align-bore. Before checking the main bearing bores, it is a good idea to use a flat honing stone dipped in solvent on the cap surfaces of the block and the mating surfaces of the bearing caps. This assures the contact areas are flat and clean. Also, chase and clean the main cap threads with a tap so torque wrench readings will be accurate.

Our cylinder walls needed more than a touch-up hone, so we decided to bore the block .030-inch over and use .030-inch over Speed-Pro forged pistons. The boring is done in two steps. First, the block is set up on a boring bar that locates the block off the main saddles so the bores are perpendicular to the crankshaft. Each bore is enlarged .025-inch on the boring bar. The last .005-inch will be done on a Sunnen CK-10 honing machine. Each piston is checked with a micrometer before honing. This way the same piston-to-wall clearance (.0035-inch in our case) can be maintained in all cylinders. The honing was done with a torque plate in order to simulate the loads imposed on the cylinders by the cylinder heads.

The block was then decked to make sure that the distance between the centerline of the crankshaft and the head surface (deck) was the same at all corners of the block. The tolerance was held to .001-inch. After the machine work was performed on the block, the block was again hot-tanked to remove honing oils and any accumulated debris.

With the block out of the way, Larry's crew moved on to the internal pieces. The crankshaft was checked with a micrometer and found to be .0005-inch out of round, so it was reground. Before being reground, the oil holes were chamfered with a hand-held grinder to improve oiling. Then the crank was reground and micropolished to

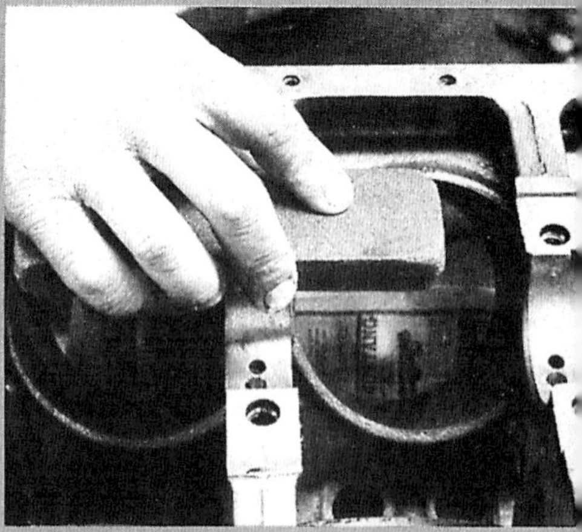

(8) All mating surfaces for the crank saddles should be touched up with a honing stone. This ensures good metal-to-metal contact and accurate torque readings.

(9) With the main saddles torqued in place, the bore of the crank bearings is checked. If they are not within tolerance, align-boring is necessary to correct the problem. Our block, like most 289 blocks, didn't need to be align-bored.

(10) A boring bar was used to enlarge each cylinder .025-inch. The boring bar locates the block so that all bores are perpendicular to the crankshaft.

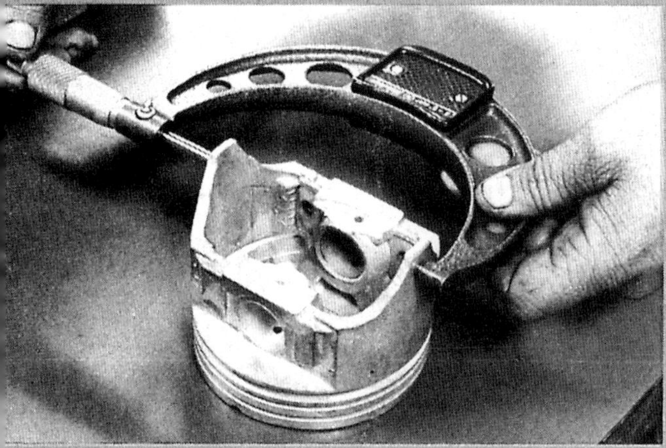

***(11)** Each Speed-Pro piston is miked before the final honing operation. This way the same piston-to-wall clearance can be maintained throughout the engine.*

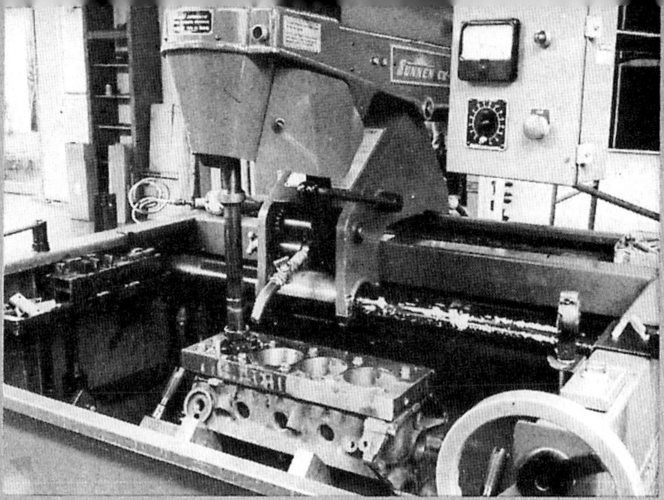

***(12)** The last .005-inch of the boring operation is performed on a Sunnen CK-10 honing machine. A torque plate is bolted to the block to simulate the cylinder head.*

make sure there were no burrs sticking up above the bearing surfaces. Micropolishing gives the crankshaft a mirror-like finish.

The next step was to rebuild the rods. First, the rods are Magnafluxed and bead-blasted clean. Then, the rods are disassembled in a rod vise, which has smooth surfaces designed not to damage rods. A honing stone is used on all mating surfaces, and the rods and caps are put in a Sunnen cap and rod grinder to make the contact areas perfect. The rods are reassembled and put on a rod hone to hone the bearing area. Then the rods are put in a rod heater that heats the small end of the rod to 500 degrees F., and the pins are pushed into the pin boss bores on a pin fixture so that the pins are centered. The rods are put in an alignment fixture so that the pins are perpendicular to the center of the bore. This ensures the piston will travel straight up and down the cylinder bore.

The rods are balanced by weighing both ends and finding the lightest rods for each end. All big ends are matched to the lightest one, as are all the small ends. The excess weight is removed from the pads at each end of the rods.

The crankshaft is balanced with the flywheel and the harmonic balancer attached because the 289 Ford is externally balanced. The crank is balanced on a Stewart-Warner balancer, which uses bobweights to simulate motion. A formula involving rotating mass (big end of rods, bearings, and a few grams of oil) and reciprocating mass (weight of rings, the small end of the rod, the wrist pin, and piston) is used to determine where to add or subtract weight on the

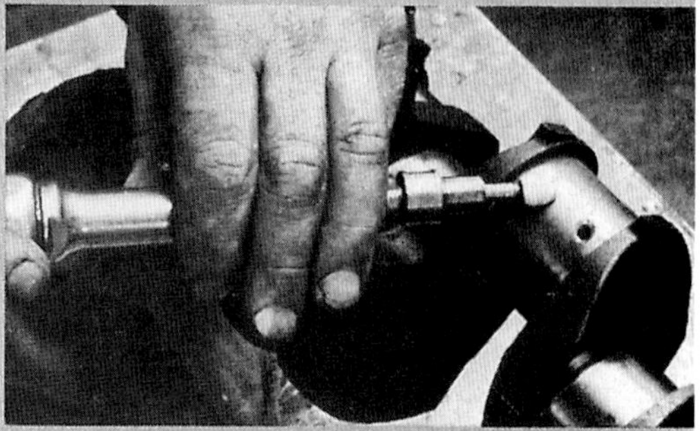

***(13)** Before the crank was reground, the oil holes were chamfered (right) with a hand-held grinder to improve oiling.*

***(14)** All the rods are Magnafluxed and rebuilt prior to balancing. Rod assembly and disassembly is done in a rod vise to keep from damaging the rod.*

***(15)** Each end of the rod (below) is weighed separately to find the lightest ones. All the rods are matched to the lightest ends.*

***(16)** To match the rod weights, the pads on the ends of the rods are ground on a belt grinder.*

BLUE-PRINTING THE 289

crank counterweights.

That just about covers the machine work on the block except checking the cam bearings. New cam bearings were installed and checked for fit with our new Crane cam. It is a rare case when align-boring is needed, but it should always be inspected anyway.

ENGINE ASSEMBLY

After all the machine work was completed, we gathered up our freshly balanced parts from Valley Head Service and took them to Edelbrock's dyno room. We also brought along a big goodie box of new aftermarket parts that Jim Mc Farland picked as the best matched components for our needs.

The first order of business was to mount the bare block on an Eagle Specialty Products heavy-duty engine stand (part No. 3001991). We converted the three-wheeled engine stand to a sturdier four-wheel model with Eagle's No. 3001992 conversion kit, which fits all three-wheeled Eagle stands. Then the engine was rolled outside and thoroughly cleaned with solvent to remove any contaminants left during the machining operations. After the solvent, the block was washed with soapy water, rinsed, and air-dried. It may seem like we are overdoing this cleaning business, but cleanliness is vital to a properly blueprinted engine.

With the block cleaned and dried, it is a good time to apply paint to the lifter galley if you feel so inclined. The main purpose of the paint is that metal fragments will show up more readily when you inspect the engine later. Also, the paint aids oil return. Curt uses Rust-Oleum Damp-Proof red primer No. 769 and lets it dry 24 hours.

The first part of the assembly process was fitting the rings. Speed-Pro R-9343, .035-inch-over rings were used so they could be individually ground to fit our .030-over bores. The rings were fitted to their respective cylinders by inserting each ring in the bore, measuring the end gap, removing the necessary amount by filing, and reinserting for a final measurement. The rings were correctly positioned in the bores by pushing them with a piece of round aluminum stock that Curt uses just for fitting rings. The top of a piston will also work for this purpose. The rings were filed down on a ring grinder, but they can also be ground with a hand-held file. The gap on the top ring was .014-inch and the middle ring gap was .012-inch. Curt makes the top ring gap larger than that of the second ring because the top rings absorb more heat, which causes them to expand more than the second rings. The ring gaps are always measured with a flat feeler gauge.

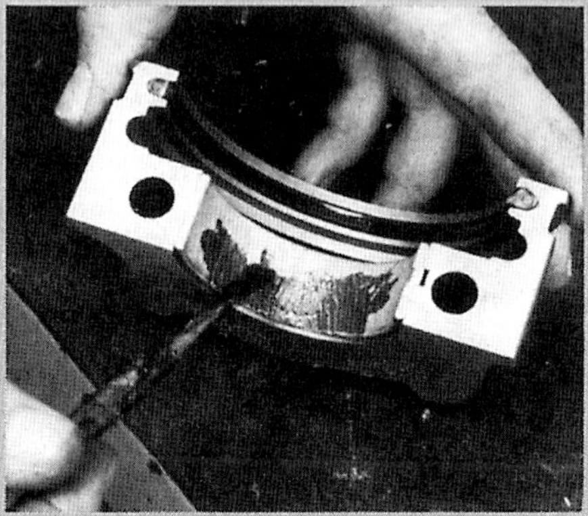

(17) Before the engine was assembled, it was thoroughly cleaned—first with a degreaser, then soapy water, and then clear water. The engine was mounted on an Eagle engine stand.

(18) All bearing surfaces were lubricated before final assembly. First apply assembly lube; then, cover the area with motor oil and work it into the bearing with your finger.

(19) Curt makes constant use of a torque wrench when blueprinting an engine. An engine stand greatly simplifies all assembly work.

Before installing the rings on the Speed-Pro .030-inch-over forged pistons (part No. 8-7006P), we cleaned the pistons, rods, and caps in solvent and dried them with an air hose. The rings can be installed by hand, but a ring expander is inexpensive insurance against breaking a ring. Put the oil rings on the piston with the expansion ring's break pointing at the center of the wrist pin. Offset the two retainer rings for the expansion ring 45 degrees to each side of the wrist pin. The second ring's break should face 180 degrees away from the oil ring, and the top ring's gap should be on the same side as the oil expansion ring.

Curt always double-checks all work done by the machine shop, just in case there ever happens to be an error. Never take anything for granted when blueprinting an engine. Check and double-check every measurement and facet of the engine-building operation. It is much easier to remedy mistakes before the engine is assembled than after. Everything on our engine was as it was supposed to be.

The main seals from a Rocket gasket set (part No. R1410) were installed, and then the crankshaft bearings (Sealed Power .010-inch-undersize) were lubricated and installed. Curt puts a light coat of assembly lube on the bearings and then covers them with 30-weight motor oil, working the lube into the bearings with his fingers. This lubrication helps protect the engine when it is first started. Then the crankshaft end play was checked with a dial indicator and found to be .005-inch, which falls right within the .004 to .008-inch range of acceptability.

The rod bearings were lubricated in the same manner as the main bearings, and then the pistons were checked to ensure they were correctly installed on the rods. The number on the rod and on the piston should face toward the outside of the engine. The rod bearings were Sealed Power .010-inch-undersize (technically oversize) units because the crank was ground .010-inch under. Before in-

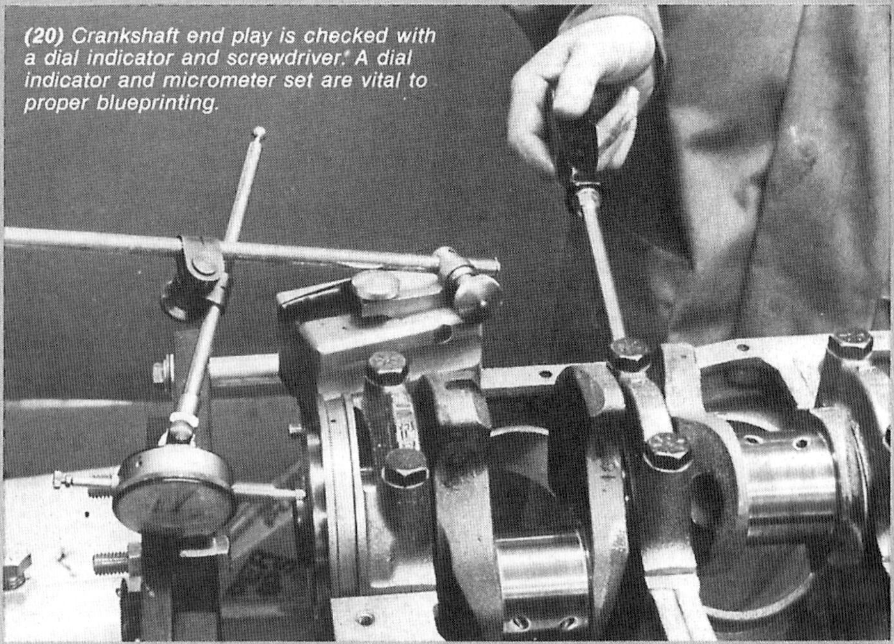

***(20)** Crankshaft end play is checked with a dial indicator and screwdriver.* A dial indicator and micrometer set are vital to proper blueprinting.*

***(21)** This is a ring-locater made by Edelbrock's Murray Jensen. A smaller disc of aluminum fits inside the bore to position the ring squarely for gap measurement.*

***(22)** Use a flat feeler gauge to measure ring end gap. A piston will work well to locate the ring for measuring. Each ring should be matched to a particular cylinder bore.*

***(23)** A ring grinder is the easy way to enlarge the amount of end gap. Speed-Pro .035-over rings were used and fitted to our .030-over block.*

***(24)** A ring expander is the safest way to install rings without breaking any, but the rings can be installed by hand. Spiral the ring around the piston as shown. The piston is a Speed-Pro forged unit.*

***(25)** A ring compressor sleeve should be used to install the pistons. Gently tap the pistons into the bores, being careful not to scratch the crankshaft.*

BLUE-PRINTING THE 289

***(26)** To be sure we had the desired compression ratio, Curt measured the combustion chambers' displacement with a burette.*

stalling the piston/rod assemblies, Curt made sure the crank spun freely and then put a generous coating of 30-weight oil on the pistons and cylinder walls. A ring compressor was used to fit the pistons, which were carefully tapped in place with a rag-covered mallet handle. The rod caps were torqued to 24 ft.-lbs.

To be sure the compression ratio was where we wanted it, Curt measured the piston and cylinder head displacement. Even though the Speed-Pro pistons were 10.5:1 units, the Valley Head Service porting job increased the cylinder head capacity, and we ended up with a 9.4:1 compression ratio. This was perfect since carbon buildup should raise the ratio after the engine is driven a while. If we wanted the 9.6:1 ratio right away, we could use a slightly thinner head gasket.

***(27)** Rod side clearance is checked with a flat feeler gauge. If there isn't the specified clearance, additional machine work is necessary.*

Next, the cam was installed, and the cam that perfectly matched our parameters turned out to be a Crane Blazer 288-2H grind (part No. 36380), designed for smooth idle and good low to mid-range horsepower and torque. The Crane Blazer cam's specs are: *intake*, 218-degree duration, .480-inch valve lift; *exhaust*, 230-degree duration, .496-inch valve lift; and a 114-degree *centerline*. We selected the Cam-Ponent Kit (part No. 36301), which included the complete cam and kit consisting of hydraulic lifters (part No. 99280), inner and outer valve springs (part No. 99858 and 99836), steel spring retainers (part No. 99944), and Crane valve seals.

Like all other parts, the cam was cleaned before being installed. The cam lobes were coated with Crane Assembly Lube (part No. 99000) to protect the cam during initial fire-up. The cam was gently worked into the block, using a screwdriver to support the outer end of the camshaft. A Crane timing gear and chain (part No. 36999) were also installed.

After installation, the cam was degreed off the lifters using a dial indicator and degree wheel. A solid lifter must be used to obtain accurate readings. Curt set our cam 2 degrees ahead of the recommended specs because the cam will retard as the timing chain stretches. With everything checked out on the cam, the front engine cover was installed. Since our original water pump had a large crack that had been filled with plastic body filler, we installed a new Sealed Power water pump (part No. PC-315). All-new freeze plugs were also installed, using a Sealed Power freeze plug kit (part No. SH-510 S). A little Loctite was applied to the freeze plugs before inserting them into the block. Remember, the cam blocks access to some of the plugs, so be sure to install them before installing the camshaft.

The last part of the short-block assembly was to install the oiling system. Since this is one of the weakest links in the 289 Ford, everything was replaced with superior aftermarket parts. A heavy-duty Sealed Power oil pump (part No. 224-411Z8) was installed with a Milodon oil pump shaft (part No. 22500). The Milodon shaft is made out of 4130 chrom-moly and features heat-treated ends for maximum strength. We installed a Milodon Super Street oil pan, which has a seven-quart capacity but a low profile for good ground clearance. A Milodon oil pickup (part No. 18375) must be used with the Super Street oil pan. We also used a set of Milodon's trick pan bolts (part No. 85010) to hold the beautiful gold-irridited pan in place and eliminate oil leaks. Normally, this would have been a good time to install the fuel pump (in our case, a Carter Super Pump, part No. M6904), but since Edelbrock's dyno has its own fuel system, we didn't install the Carter pump until after the dyno tests.

THE CYLINDER HEADS

With the short-block assembled, we turned our attention to the cylinder heads. The valve job and head porting were also handled by Valley Head Service. Ford fanatics will be interested to know that Valley Head Service did all the competition headwork for Carroll Shelby. Our engine received a "Marathon" valve job and a "Streetmaster" porting treatment.

First, the heads were hot-tanked

and Magnafluxed. Then the stock valve guides were bored out and replacement guides were pressed in place and spot-faced to the stock valve guide height. Sealed Power stock-size replacement valves were used, and each was matched to its guide after the stem was checked to guide clearance.

The next step was the Streetmaster porting job, basically a competition porting job for the street. The difference between a full competition job and the Streetmaster is that the Streetmaster doesn't place as much emphasis on the intake ports. After 1000 miles or so, the intakes get loaded up with carbon, so a lot of porting here is wasted money. Templates are used as a guide to how much metal to remove, and then the final touches are done by polishing the ports.

The Marathon valve job is a competition-type valve job for street applications. The exhaust seats are ground to an exact width of .080-inch, and the intake seats are ground to .060-inch. All seats are located .0015-inch from the edge of the valves. The intake valves receive a three-angle grinding of 45, 35, and 25 degrees. The exhaust valves receive a two-angle grinding of 45 degrees with a 25-degree backcut. Then the valves are hand-lapped to ensure a positive seal and verify proper valve location.

Following the valve job and porting, the heads were assembled using new Sealed Power rocker arms (stock-type replacement units) and Crane double springs, retainers, and seals. Spring shims were used to ensure installed heights within ±.005-inch. All springs were checked for binding.

The Rocket head gaskets were installed and the heads torqued in

(28) There are several ways to degree a cam. This is the lifter method. Be sure to use a solid lifter for accurate readings, even though you may have a hydraulic-lifter cam.

(29) The oiling system needed a little help (above right), so we installed a new Sealed Power pump, Milodon heavy-duty shaft, and Milodon pickup.

(30) Take time and care when inserting the camshaft. A screwdriver works well to support the end of our Crane Blazer cam. Lubricate the cam thoroughly with Crane Assembly Lube.

(31) The special Milodon oil pickup is necessary for use with a Milodon Super Street seven-quart gold-irridited pan. The pan features a low-profile for street use.

BLUE-PRINTING THE 289

place, using the pattern specified by Ford, to 70 ft.-lbs. The Crane hydraulic lifters were slipped in place after being thoroughly covered with assembly lube while the lifter bores were covered with oil before the lifters were installed. The pushrods were also oiled, and the pushrod ends were covered with assembly lube before being in-

(32) A Streetmaster porting job starts with a template to scribe the rough shape where material will be removed from the head. The ports are then ground and finally polished.

(33) All areas of the head that are worked over are carefully polished before the job is finished. Old valves (not the new ones) are used during the grinding and polishing procedures.

(34) A Valley Head Service Marathon valve job includes a three-angle grinding on the intakes and a two-angle grinding on the exhausts.

(35) After the machine grinding, the valves are hand-lapped to their final fit.

(36) The factory valve guides are drilled out and replaced with new pressed-in guides. The new guides are honed to fit each valve stem.

***(37)** All the valvetrain pieces are laid out and checked prior to assembly.*

stalled. Stock Ford pushrods were used in conjunction with Sealed Power rocker arms (part No. 2R-836). All contact areas were coated with assembly lube. Curt likes to put plenty of oil in the center of the rocker arms so they have oil when the engine is first fired. Valve lash was set at zero because of the hydraulic cam.

The final touch was to cover the valvetrain with a pair of polished Edelbrock valve covers (part No. 4160). Curt painted the area between the fins with Rust-Oleum National Blue paint to match the block. The intake manifold best suited for our needs turned out to be an Edelbrock Torker (part No. 2755). The Torker is a high-performance street manifold designed for midrange rpm, which is exactly where our engine will operate best.

***(38)** The Crane springs were checked with a caliper and spring tester prior to being installed on the engine. Spring shims were used to keep installed heights within ± .005-inch.*

***(39)** The Crane hydraulic lifters were thoroughly lubricated before being installed in the lifter bores. The pushrods and rocker arms were also generously covered with oil and assembly lube.*

The recommended carburetor is a Carter 9627 Super Quad AFB, which is rated at 625 cfm and has an electric choke for easy starting.

The entire ignition system is comprised of Accel parts, including a BEI-II breakerless ignition (part No. 39201), 9mm "Fat Stuff II" yellow silicone supression-core spark plug wires, Yellow Jacket spark plugs, and a Super Coil (part No. 140001). If someone wanted to use mechanical points, a good choice would be one of Accel's dual-point distributors. Dual-point distributors were used successfully on the Shelby 289 engines. The rest of the electrical system—alternator (part No. 190201), regulator (part No. 200201), and starter solenoid (part No. 40200)—is also from Accel. The starter is a Weber Ultra-Tork unit (part No. 397102).

The headers are Eagle Gold Label headers (part No. 2101) from Eagle Specialty Products, Inc. The Gold Label headers are made of extra-strong, 14-gauge .075-inch steel and coated with baked-on gold paint. To keep the engine properly lubricated, a Pennzoil PZ-1 oil filter was installed and the oil pan filled with seven quarts of Pennzoil 20W-40 oil. The engine was removed from the engine stand and installed on Edelbrock's dyno with a McLeod flywheel and bellhousing.

The engine sounded great during the break-in cycle, and everything checked out fine; so Curt and Edelbrock's Murray Jensen proceeded with the power runs. The total lead was set at 36 degrees, the carburetor was left in its out-of-the-box state, and H&H racing gas was used. The engine ran strong all the way to 6000 rpm, our imposed limit. With the stock rods, caps, and bolts, we were hesitant to push the engine any farther, although the power curve was still going up at 6000 rpm. Our final figure at 6000 rpm was 276 horsepower with 242 ft.-lbs. of torque.

DYNO RESULTS

RPM	HORSEPOWER	TORQUE (ft.-lbs.)
2000	79	207
2500	103	217
3000	131	229
3500	168	252
4000	197	258
4500	226	264
5000	252	264
5500	275	262
6000	276	242

BLUE-PRINTING THE 289

The final figure was short of our original goal of 300 horsepower, but the results still indicate that we will have a great street engine. The torque curve was amazingly flat, which means that all the engine pieces are well-integrated and the engine is very efficient. According to Edelbrock's Jim McFarland, a flat, broad torque curve is the sign of a strong street engine. The closeness of the torque figures indicates that very little additional effort is needed when the engine moves up to a higher rpm range. This also means that the engine is very fuel-efficient and should easily provide 15 to 20-mile-per-gallon economy. Jim said we probably could have experimented with tuning techniques on the dyno run and squeezed another 15 horsepower out of the engine, but our tight time schedule precluded that.

Jim also pointed out that our 276 horsepower was *honest* horsepower, whereas many factory ratings are optimistically unrealistic. As a case in point, he stated that the Mustang high-performance 289, which was rated at 271 horsepower, probably put out 230 horsepower at best, and without the efficiency of our blueprinted engine. So, all things considered, we were very pleased with the results of our 289 Ford engine blueprinting operation. **HR**

***(40)** The finished engine as hooked up to Edelbrock's dyno. A McLeod flywheel and bellhousing were used to mate the 289 Ford to the dyno. The engine ran like a champ during the tests.*

***(41)** The induction system consists of a Carter 625-cfm carburetor and an Edelbrock Torker manifold. Also visible in the picture are the Accel BEI-II breakerless ignition and the Edelbrock polished valve covers.*

***(42)** The finished engine waiting to be installed in our '65 Mustang. With all the blue paint, gold Milodon oil pan, gold Eagle headers, yellow Accel ignition pieces, polished Edelbrock valve covers, blue Chromatized Rocket air cleaner, fan, and valve cover wing nuts, and stainless steel-braided hoses and Econo-Seal hose ends from Earl's Supply, the small-block Ford looks as good as it runs.*

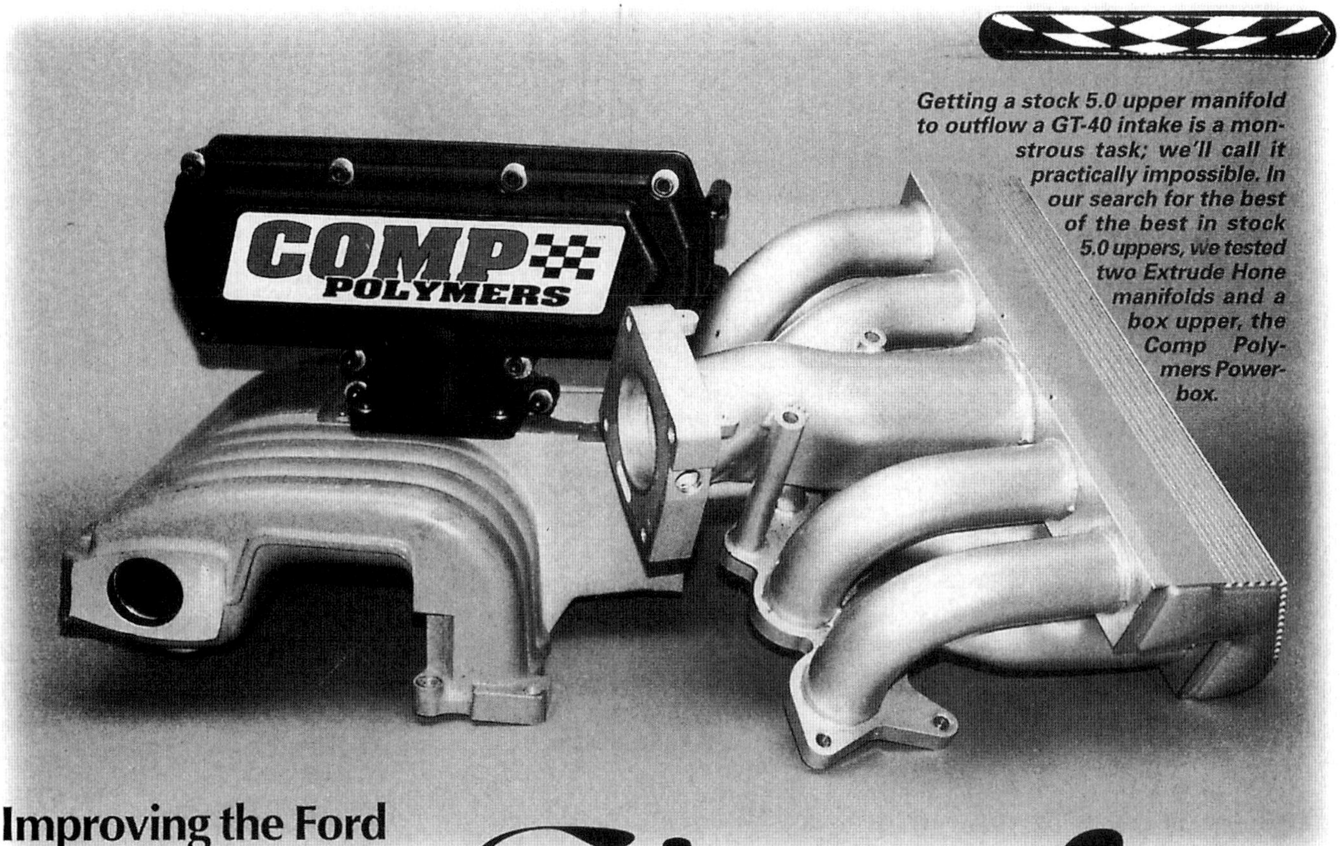

Getting a stock 5.0 upper manifold to outflow a GT-40 intake is a monstrous task; we'll call it practically impossible. In our search for the best of the best in stock 5.0 uppers, we tested two Extrude Hone manifolds and a box upper, the Comp Polymers Powerbox.

Improving the Ford 5.0 Upper Manifold

Stock Intake SHOOTOUT

By James Lawrence

It's hard to believe Ford began engineering the 5.0 Fuel Injection intake manifold more than 14 years ago. For those of you who aren't mathematically inclined, this means the average 21-year-old 5.0L enthusiast was 7 years old when engineer Wally Beaber of Ford's Powertrain division sat down at his coffee table and began laying out plans for the future of late-model Mustang performance.

Motorsport's GT-40 intake (left) is hard to beat because of its shorter runners and larger plenum. Compare for yourself against the stock intake. Even after hand porting or Extrude Honing the stock manifold, it's hard to work around this problem.

Unfortunately, at the time Beaber was busy trying to make the best design possible for the factory 302 EFI intake, he had to think about goals other than making maximum horsepower. To a production vehicle like the 5.0 'Stang, important details are cosmetic packaging, space considerations, and low-speed torque requirements. Beaber told *5.0* that in the design process, more than 27 combinations of tubing length and diameter were tried in the effort to develop the best intake manifold that would satisfy all of Ford's requirements.

Fourteen years later, enthusiasts are making 500-plus hp using the timeless design that Beaber engineered in late 1984. We covered improving the stock lower intake in the Apr. '98 issue, so in this segment, we are concentrating on the more problematic upper manifolding. The biggest problems with the upper are the length and size of the intake runners, the small size of the air plenum, and the intake's restricted inlet.

The factory upper manifold runners are very long and very small, which doesn't help high-rpm power any. The long runners become a restriction on any engine that is revved into the 5,500- to 6,000-rpm range. Since that is the point at which most modified 5.0s get shifted, we have a problem.

There is no easy way to decrease the runner length on a factory manifold other than cutting open the plenum area and milling out the runners in a machine shop.

Although this was common in the early days of 5.0 performance, it's no longer cost-effective or even practical. Increasing the diameter and volume of the runners is possible, however, and it helps raise the usable rpm band. But if you are trying to port the intake manifold, how do you get inside to increase the size of the runners? If you are using conventional porting technology (carbide bits and a grinder), you must cut open the intake, drastically increasing the cost and effort involved.

If you refer to the first part of this shootout, you'll recall that we had two reputable hand-porting shops, Competition Heads and Fox Lake Power Products, do the grinding on a pair of lower intake manifolds. But these shops told us they weren't interested in porting an upper manifold for our test. As Ron Robart of Fox Lake explained, "Porting the upper manifold, cutting it up, or boxing the plenum to add volume is a waste of time compared to the cost involved." Robart estimated that he would have to charge $800 to work a stock upper manifold fully and make it flow enough to match one of his ported lowers.

So what is one to do? If one is committed to retaining the stock upper manifold, the most effective solution so far seems to be the Extrude Hone Powerflow porting process, which we also showcased in our previous installment. During the process, an abrasive media is forced through the manifold, removing material where the greatest restrictions exist. Extrude Hone is perfect for the stock manifold application, because it does not require cutting apart the manifold, and it can increase the volume and flow of the runners. Also, it can reduce any airflow restrictions such as rough casting marks.

While looking at the top of the stock manifold with the cover removed, you can see that Ford kinked the air inlet to make room for the upper-to-lower intake bolts. This kink is restrictive, not only for that reason, but also because the stock casting is small in this area and has plenty of room for porting. In fact, the stock manifold's entry is only sized for the stock 60mm throttle body size. Either through hand porting (which is

Wow! That's what the 5.0 staff members said when we saw the dramatic runner length differences illustrated in this photo. The stock upper manifold has a runner of approximately 14 inches compared to 1 inch for the Comp Powerbox. The shorter runners will help high-rpm power a whole bunch on naturally aspirated engines and will also kick some ass on blower applications. For mild street cars, if you bolt on a short-runner box-type intake, torque will suffer.

Stock Upper & GT-40

The two legendary Ford 5.0 manifolds stand side by side. The weak factory piece can't hold a candle to the Ford Motorsport GT-40 upper. The GT-40 offers larger runners and a plenum designed around a staggered runner concept, rather than the stock intake's inline runners. It's not surprising that the GT-40's runners flow almost 100 cfm over stock.

Stock Upper Versus GT-40

Stock Upper:

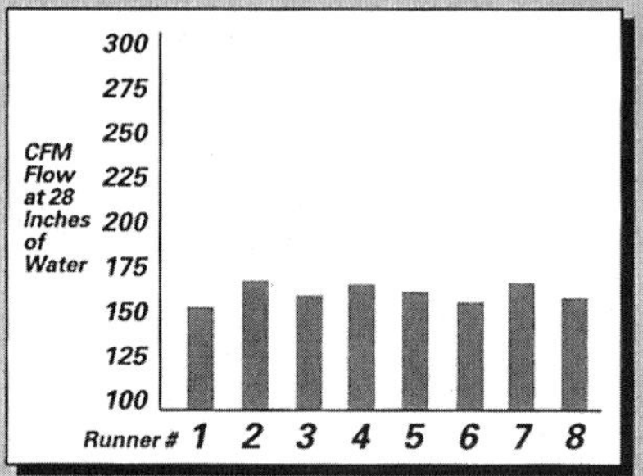

GT-40

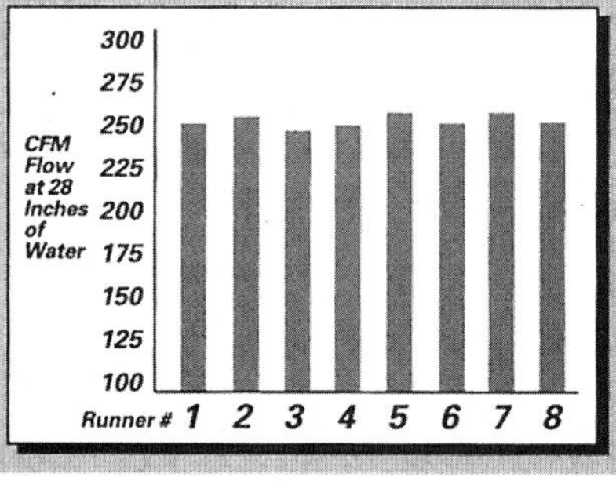

Extrude Hone

Extude Hone sent us two different ported upper intake manifolds for our shootout. Extrude Hone offers different levels of abrasive honing, from street use to more radical applications, to satisfy every customer. For our testing, we received its standout upper manifold as well as a max-effort stock intake that was ported to the very limits of the casting.

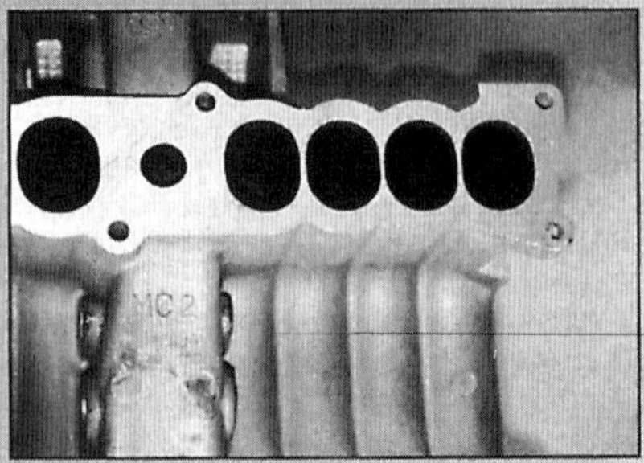

Another neat thing about the Powerflow process is that you can specify any throttle body size opening (from 65 mm to 75 mm) and you receive a glasslike port finish that is great for port velocity.

Extrude Hone I Versus II

Extrude Hone I

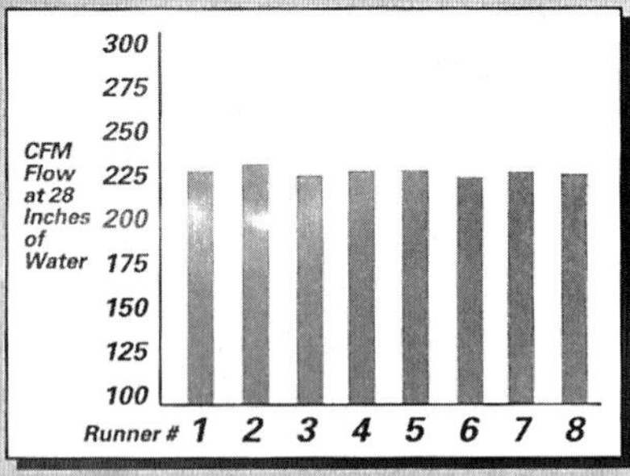

Extrude Hone II

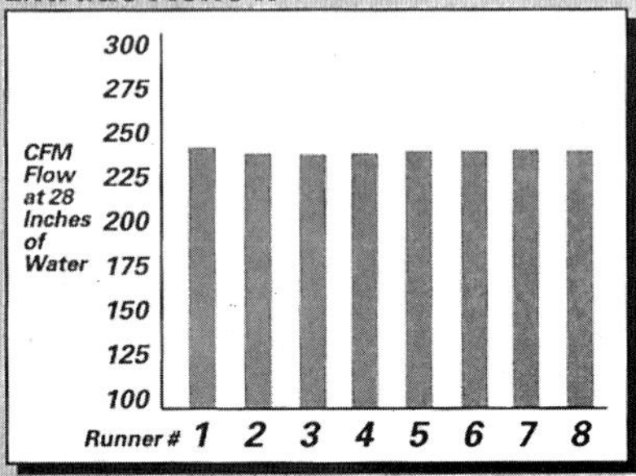

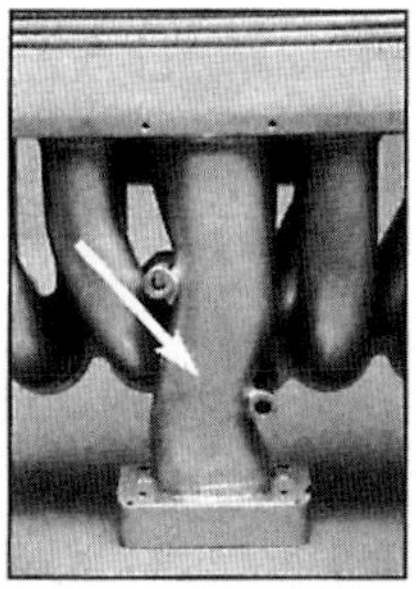

The kinked inlet runner of the factory manifold (left, arrow) restricts airflow significantly. Wally Beaber of Ford's Powertrain division told us that it was required for mass-production purposes. The GT-40 intake still has a kink (center, arrow), but the air inlet to the plenum is much larger and the kink is less severe. 5.0 racers have been modifying the manifold by taking the kink out of the GT-40 upper to increase flow, but it doesn't help much. Comp's Powerbox inlet (right, arrow) is completely unrestricted and no doubt helped this intake manifold to come out No. 1 in flow.

Comp's Powerbox has an inlet sized for only a 65mm throttle body, but it can easily be opened up to a larger size. Looking through the inlet shows that there is no restriction present here, just pure massive direct airflow pouring into the short runners. It's no surprise the Powerbox's runners flowed almost 270 cfm.

fairly difficult) or through Extrude Honing, you can increase the size of the air inlet, making this area much less restrictive. In fact, with the Powerflow process, you can specify your exact throttle body size, and your manifold opening (and innards) will be opened to the requested size.

Comp Polymers, a division of Competition Cams, offers a replacement upper plenum called the 5.0 Powerbox that corrects many of the problems inherent in the original Ford design. Constructed of polymer-composite material, the Powerbox was designed from a clean sheet of paper with higher-rpm horsepower in mind. Recall what we wrote earlier about the runner lengths being a restriction in the original EFI manifold. The Comp engineers already knew about this and designed the Powerbox as a short-runner or box-type upper manifold and increased the plenum size significantly.

The Powerbox has additional features that allow it to be a cut above a stock manifold's design. It's a modular two-piece, bolt-together design, so the air inlet of the manifold is unrestricted. To bolt the upper manifold to the lower, simply unbolt the entire upper intake, and the bolts are housed within the upper plenum. As for the air inlet opening, the manifold is supplied with a 65mm throttle body opening, but it can be opened up easily to as large as 80 mm if desired.

The short-runner versus long-runner argument is extensive, but it boils down to whether you are more interested in a broad torque band or maximum horsepower and rpm potential. After launch in a drag-race or street/strip application, your 5.0's engine will rarely dip below 4,000 rpm, and since the short-runner manifold will make most of its power in the 4,000- to 6,000-rpm range, it will accelerate more quickly and make more peak power than an intake with a longer runner. In addition, the larger plenum of the 5.0 Powerbox will be better able to fill those shorter runners with air in the upper rpm range. One other thing to think about is this: The Powerbox will work perfectly with a blower or nitrous combination where loss of torque isn't a consideration, and the short runners will really allow the motor to breathe to its maximum.

For street use, when a stock camshaft is used and the shift rpm of the engine is 5,500 rpm or lower, or when you are running a tall gear ratio (3.73 or taller), the short-runner manifold will be a disappointing addition.

The loss of torque isn't necessarily a bad thing, especially if you have a street/strip or street-race vehicle and you have a problem hooking up your 5.0 on street tires. The better top end of the box manifold (assuming you have the gear to put the engine in the higher powerband) will allow you to avoid tirespin off the line and have the motor really scream higher in the rpm range. But, as mentioned earlier, you need a camshaft that will work within the powerband (6,000-rpm minimum shift point), and the car must be geared to help compensate for the loss of torque. Everything's a compromise.

The best situation for a street vehicle is to have a medium-length runner that has enough volume and plenum volume to support a 6,000-rpm shift point, and thus you can see why aftermarket manifolds such as the SVO GT-40 and Edelbrock Performer 5.0 are so popular. But still, flying with stock parts is cool, cost-effective, and highly possible. That's why we wrote this article. So next time you get your doors blown off by some guy running a blower and a stock-*looking* manifold, don't say we didn't warn you.

Comp Powerbox

The Powerbox by Comp Polymers is an excellent manifold for high-rpm naturally aspirated or blown 5.0s. Short runner length, a generous plenum, and a direct, efficient air inlet are the greatest features of the Powerbox. We at *5.0* feel that this manifold, with a welded and ported lower intake and in a blower application, could run as quick as the mid to low 10s or make as much as 550 hp. If you've got a grocery getter or if you sit in traffic, this manifold's lack of runner length might not make it your best choice.

Comp Powerbox

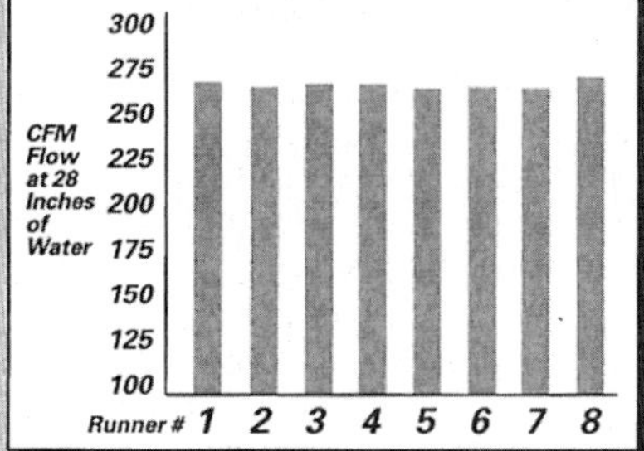

Sources

Comp Polymers
Dept. 5.0
3406 Democrat Rd.
Memphis, TN 38118
800/999-0853

Extrude Hone
Dept. 5.0
8800 Somerset Blvd.
Paramount, CA 90723
310/531-2976

Fox Lake
Dept. 5.0
6512 Dalton Fox Lake Rd.
N. Lawrence, OH 44666
330/682-8800

Street POWER Combos

Small-Block FORD

By Pat Ganahl

We're about to open a can of worms. That's inevitable when the bench racing discussion turns to building a hot small-block Ford.

For starters, what exactly *is* a small-block Ford? There are at least four possible versions, all more or less related. First is the 289-302 V8, which is definitely a *small* block. Second is its big brother, the 351-Windsor, which has a taller deck and much beefier bottom end with massive 3-inch main journals. Both the 289/302 and 351-W engines are considered "Windsors," and use similar, inline valve wedge heads. Third, we briefly had the Boss 302 ('69-'70) which combined the 302 short-block with Cleveland 4-barrel heads. Fourth is the 351-Cleveland, which uses a completely different block and a cylinder head with canted valves and huge ports. (A fifth version, the 351 and 400 "M" engines, uses Cleveland heads and a Cleveland-type block with approximately 1-inch-taller decks and larger crank journals.) Although nothing else interchanges between Windsor and Cleveland engines, the heads will (if water outlet passages at the front are drilled and plugged appropriately; Cleveland water outlet is through the block, while the Windsor passes coolant through the intake manifold). To further complicate the issue, Ford Motorsports has recently released a 351-W SVO high-performance block which uses the Cleveland crank and comes in Windsor (9.500-inch) or Cleveland (9.200-inch) deck heights.

We can't possibly explain the nuances of all the potential head/block/manifold combinations in this limited space. For starters, I would *strongly* suggest you obtain a copy of *Ford High Performance*, which not only shows how the various combinations go together, but also contains the complete Ford SVO Motorsports catalog and a list of dealers. It's available at SVO Ford dealers, or by mail from: I.M.I., Dept. HR, P.O. Box 26226, Birmingham, AL 35226 ($5 plus $1.75 shipping).

HEAD GAMES FOR HORSEPOWER

To come up with our mild and wild horsepower recipes, we're going to have to do some "combination" juggling. We realize that what you want to build probably depends largely on what you already own. But with Fords there are definite limitations. In 9:1 streetable form, don't even *dream* of making 350-plus horses with a 302, or 400 ponies from a 351-W, using stock-type heads. The big problem with Ford small-blocks is that the Windsor heads (especially the 302s) are generally too small, and the Cleveland heads are too big. The ports and valves in Boss 302 and 4-barrel Cleveland heads are ridiculously large (one reason why they haven't been made since '73). The 2-barrel heads, which have much smaller—though still very generous—ports and 2.04-inch/1.65-inch valves, would make an excellent low-buck street-performance combination, except

It takes some headwork, but this small, innocent-looking Ford 302 package is capable of 330-plus horsepower. New factory aluminum valve covers are clean.

that their fully open combustion chamber design affords no "squish" area at all, and thus they're prone to pinging on pump gas, even at 9:1 compression. So, for the sake of variety while living with limits, we'll look at stock-head 302 and 351 Windsor "recipes" for a mild 300 to 325 horses, and a 351-W or (preferably) 351-C with aftermarket heads for a wild 400 to 425 horses.

ON THE MILD SIDE

Getting an honest, streetable, 300 hp out of a stock-head 9:1 302 Ford isn't easy. "There's no way you can make a Windsor-type head make horsepower in a bolt-on condition," says Ford wizard Chris Kaufmann. For our mild 302 recipe we relied on surprisingly similar combinations built by Kaufmann and by Edelbrock for R&D research. Edelbrock's 9.4:1 "Torker" test 302 made 330 hp at 5500, while Kaufmann's "Stage III" 9.5:1 motor makes 325 to 345 hp depending on exact components. Both require porting of the Windsor heads.

Here are the basics. The weak link in the 302 short-block is the 5/16 rod bolt; if the engine will see 6500-plus rpm, we recommend high-strength replacement bolts from ARP, SPS, or B&B. A good, forged, flat-top replacement piston is the TRW L2482. Kaufmann has tested numerous cams, and likes the Edelbrock Torker Plus PN 5022 (212 I/224 E; .472/.496 lift; 112 lobes), the Comp Cams 292-H PN 31-330-3 (244 I&E; .518 lift; 110 lobes), or his own KPI Stage III PN 2522 (230 I/244 E; .535 lift; 110 lobes).

302/351W FORD SMALL-BLOCKS

	MILD		WILD	
	302	**351W**	**351W**	**351C**
CARBURETOR: Size Type Brand	 600-750 #4776-4779 Mech. Holley	 700-750 #4779, 8156 Mech. Holley	 750 #4779, 9379 Mech. Holley	 750 #4779, 9379 Mech. Holley
INTAKE MANIFOLD: Type Brand	 Single-plane Edelbrock Torker II #5021 Offy #6071 Holley #300-6	 Same Edelbrock Torker II #5081 Holley #300-18	 Same Edelbrock Torker II #5081	 Same Ford SVO #M-9424- B351
CAMSHAFT: Type Duration @ .050 Valve lift Lobe Separation	 Hydraulic I.214-234 degrees E.224-244 degrees .490-.530 inch 110-112 degrees	 Same I.224-244 degrees E.230-244 degrees Same Same	 Same I.240-250 degrees E.240-250 degrees .500-.520 inch Same	 Same I.208-220 degrees E.218-230 degrees Same Same
HEADERS: Type Primary Tube dia. Collector dia.	 4 into 1 or Tri-Y 1.5 or 1⅝ inches 2.0-2.5 inches	 4 into 1 1⅝ inches 2.5 inches	 Same 1¾ inches 3.0 inches	 Same Same Same
HEADS: Type Intake valve dia. Exhaust valve dia. Combustion chamber size	 '70 351W or '83-up 302 ported 1.84-1.90 inches 1.60 inches 62-66cc	 Same Same Same Same	 AR Pro-1 unported 1.94-2.02 inches 1.60 inches 58cc	 Ford SVO #M-6049- A3 unported 2.190 inches 1.710 inches 62cc
COMPRESSION RATIO: Compression Piston type Brand/part No.	 9.0-9.5:1 Stock or forged flat-top TRW #L2482 or equivalent	 Same Dished 23cc TRW #L2413	 9.3:1 Dished 13-17cc TRW #L2378 or #L2446	 9.5:1 Forged flat-top TRW #L2379
IGNITION: Type	 Point, electronic,	 Same	 Same	 Same

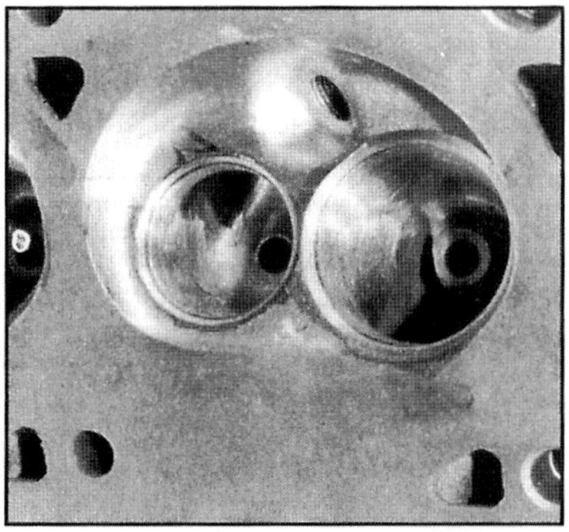

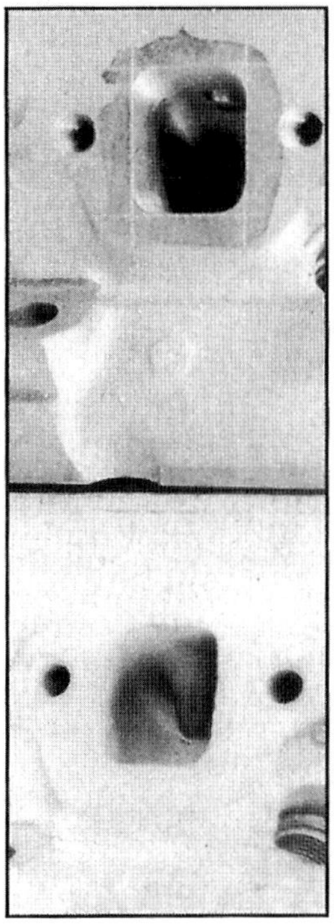

Take a close look at the chamber and port work Kaufmann does on Windsor heads to get power. Cleaning and blending the exhaust bowl is critical; open up exhaust port at outlet. Intake side of chamber is left rough to reduce air/fuel separation.

With performance cams, Fords should be converted to adjustable rockers. On '79-up Windsor heads, you can use Kaufmann 5/16-thread studs, "rail" tip roller rockers from Comp Cams, and KPI dual springs for a bolt-on combo. Other heads need machining.

There are two ways to go on heads. The best bolt-on stock head is the 1970 351-W (C9OE casting) which has 1.84/1.54-inch valves and 62-64cc chambers. However, this head requires machining of the press-in rocker stud pedestals to install screw-in studs (for adjustable rockers), and still needs a larger exhaust valve, exhaust port work, and bowl clean-up to make the numbers we want. If you have to port the heads anyway, you might as well start with more plentiful 1983-up 66cc 302 heads, with '85s and '87s being best (E5TZ casting). Do *not* use the oddball '86 "swirl" head (E6SE casting) which has a highly shrouded exhaust valve. All '79-up heads come with machined rocker pedestals drilled and tapped 5/16 inch, and Kaufmann offers 5/16-shank screw-in studs to make the conversion to adjustable rockers much simpler.

Enlarge the valve seats for 1.90-inch intakes and 1.60-inch exhausts (TRW PN V2609 and S2608), smooth and blend the transition from the exhaust seat to the bowl, paying particular attention to the first ¾ inch below the seat, put a 30/45/65-degree cut on the seat, and match the exhaust port to a 302 or 351-W gasket, depending on which size flange your headers have. The exhaust is the side that needs work in the Windsor heads; you can leave the intake alone other than blending the bowl. For springs, you want about 115 pounds on the seat and 285 pounds open at .500-inch lift. For performance, you should convert Windsor heads to adjustable rocker arms. With his own screw-in studs, Kaufmann uses the new self-aligning Comp Cams "rail" roller rocker, which makes a slick, bolt-on conversion for 1979-up heads requiring no machining, guide plates, or hardened pushrods.

Edelbrock produced their 330 horses with a No. 4776 600-cfm Holley, while Kaufmann likes the 750-cfm No. 8156 competition model. Both strongly recommend a mechanical secondary double-pumper, and state that the small Ford likes to be slightly over-carbed. Both used the Edelbrock Torker II manifold, PN 5021. For headers, you'll have to take what you can get. Hedman Huslers PN 85604 ('79-up Mustang) are excellent with 1¾-inch primaries, 32 inches long, and a flat, 11-inch-long collector. Most others have 1⅝-inch primaries, 28-30 inches long.

Wild horsepower takes non-stock heads. SVO Australian Cleveland head (left) ***combines 2-barrel ports with 4-barrel closed chamber, but not many remain. SVO aluminum high-port heads*** (center) ***are expensive, but work great. AR head*** (right) ***is Windsor design, has healthy ports, and takes Chevy valves.***

OPTIONS, OPTIONS

Now consider, if you're looking for a mild 300 horses, a multitude of options that might be simpler and cheaper than modifying the 302. First, you could bolt in a 351-W instead. The *stock* 1970 4-barrel version, at 11:1 compression, made 300 horses (10 more than the Boss 302). A 9:1 351-W with the carb, manifold, cam, and headers outlined here would be in that ballpark. With ported heads, Edelbrock got 350 hp in their 351-W tests. Second, you could bolt stock 351-C 2-barrel heads on a 351-W with flat-tops (9:1) using a B&A manifold, and make about 350 hp with similar parts. Or you can go with AR or SVO aluminum heads on the 302 or 351-W for big gains for big bucks. Using a stock 1985 302 short-block with unported AR heads (9.2:1 compression), 1.94/1.60 valves, a 292-H Magnum Comp Cam, Torker II intake with a 750 Holley double-pumper, and 1¾-inch headers, Alan Root got 380 hp on pump gas on Ken Duttweiler's dyno. With an identical combination in a 351-Windsor fitted with TRW L2378 dished pistons (9.3:1), he pulled 430 horses at 5800 rpm.

Although ports are the same, taller-deck 351-W requires a wider manifold (top) ***than 302. Dual-plane is best below 4500 rpm; single-plane, such as Edelbrock Torker, gets bigger numbers.***

ON THE WILD SIDE

This brings us directly to the "big" 400-425-hp recipe. In streetable form, you can crank out this kind of power with a 351-W or a 351-C, but you're not going to make it with stock heads. The 1971 Boss 351-C, even at 11.1:1 compression, made only 330 hp—with a very soggy bottom end due to the huge, low-velocity ports and valves. The stock 2-barrel Cleveland heads have better ports and valves but detonation-prone open chambers. A great compromise is an Australian iron Cleveland head offered by Ford SVO for about $200 each (M-6049-C351) which combines the 4-barrel closed chambers with the 2-barrel ports. They can be used on a Cleveland with Edelbrock (No. 2750), Holley (No. 300-12), Offenhauser (No. 6122), or Weiand (No. 7516) manifolds, or adapted to a 351-W by plugging and drilling water passages and using a B&A manifold. Unfortunately, however, these heads are in limited supply.

But to get over 400 streetable, 9:1, basi-

Junkyard 2-barrel Cleveland heads are worth about 50 hp over stock Windsors. B&A makes manifolds for the swap on 302s and 351-Ws.

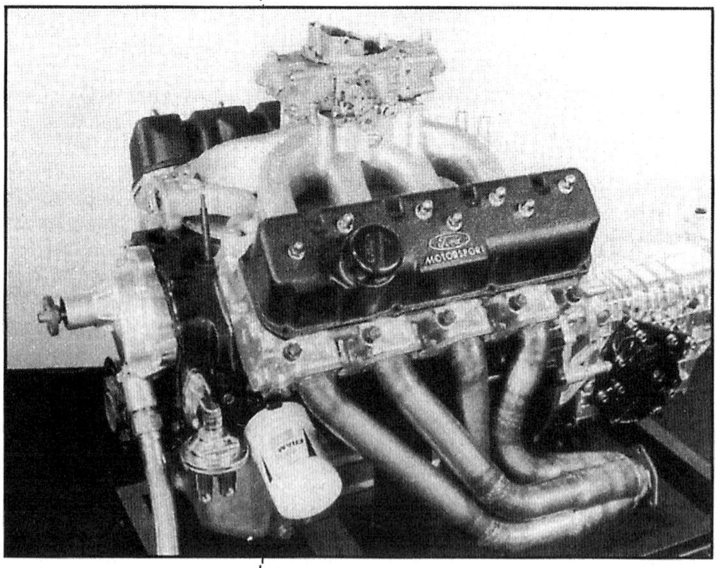

SVO Cleveland heads and manifolds are made for SVO Windsor-style racing block, but can be bolted to 351-C production blocks with minor modifications. Hall Pantera headers are shown on this mock-up, with "C351" manifold.

Continued on page 106

COVER STORY

HIGH-TORQUE STREET ENGINES

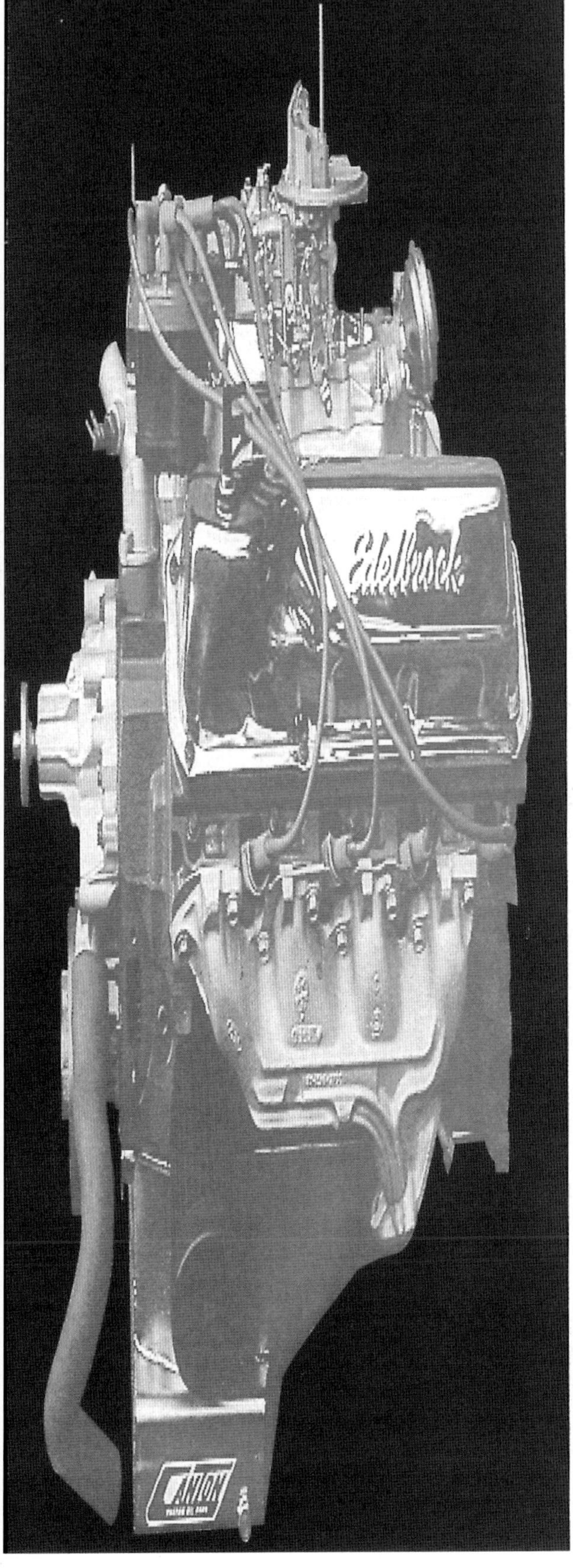

DOWN & DIRTY

The 400M Ford: 468 Lb-Ft, 380 HP for Less Than $2,000

By John Newman

We're going to demonstrate how to build a cheap 406ci Ford that runs on pump gasoline. We'll also show you how to save money by skipping steps that aren't critical to the success of a 5,200-rpm street thumper. Our scratchy ol' 406 makes 380 horsepower and at least 450 lb-ft of torque—for less than $2,000.

First, the Basics

A 406-inch Windsor, based on the 351ci Ford, is light for its displacement but costs a lot of money to build and involves a stroker crankshaft assembly. For our thumper, we used a 400M block. The 351M/400 engine is nothing like a 351 Windsor. The bellhousing bolt pattern is the same as that in a 429/460 big-block, while the engine design is basically a Cleveland small-block. These yeoman-duty motors never had more than 8.0:1 compression, and all were smog engines. Hence, there are thousands of them awaiting revival in salvage yards.

The 351M was produced between 1974 and 1979 in passcars, while the high-torque, low-rpm 400 prodded large cars and pickups through 1981. When new, these slugs never

Photography: John Newman

Since our street thumper won't ever see more than 5,200 rpm, we cinched the crank down with the stock two-bolt mains and ½-inch ARP studs.

Sheldust Crankshaft & Engine turned the 400M crankshaft 0.010/0.010 inch, leaving the standard fillet radius on either side of the journals. Sheldust also chamfered the oil holes and polished the journals.

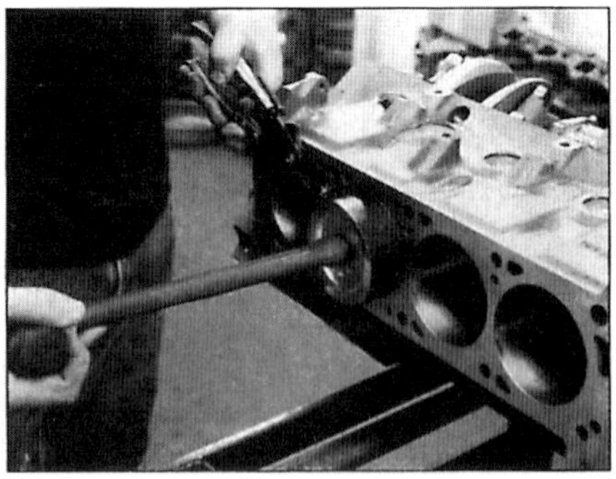

With a 0.303-inch overbore and 0.025 inch removed from the cylinder head surface, the deep-dish Silv-O-Lite pistons produce a static compression ratio of 9.45:1.

produced more than 173 hp. Regardless, "M"-series engines are plentiful, cheap, and usually show little cylinder wear, even after 150,000 miles.

If cost is a major concern, as it was with our project, then the 351M/400 engine family is the best place to begin. In addition, we adhered to three important rules while building our torquer:

1) Don't overcam. Although a cam with 330-degree duration may make the engine sound great, it won't make much torque down low in the powerband. We used a 268-duration cam, which gave us a flat powerband and grunt from 2,000 to 5,200 rpm that pulls like a turbine.

2) Don't overport. Total airflow at high rpm is not even a consideration—the idea is to keep the port velocity high. For a 351M/400 combination, use the OEM two-barrel casting, which features an open-chamber design and flows well. Match the gaskets to the heads and keep the port velocity up by using a low-rise dual-plane manifold. Less is more in port size, as long as the ports can flow enough to meet the demands of engine speed up to 5,500 rpm.

3) Don't overcarb. If you choose a four-barrel carburetor, make sure it's equipped with vacuum secondaries and flows approximately 650 cfm. If emissions or class restrictions dictate, then a two-barrel that flows 500-600 cfm will suffice.

We came up with the 400M combination after our conversation with Barrie Poole. The "Border Bandit" out of Sandy Elliott Ford in Windsor, Ontario, Canada, was a dominant NHRA Super Stock racer in the late '60s and early '70s. What follows is Poole's combination, updated with 1998 camshaft technology.

Our nest egg was a 135,000-mile wheezer worthy of 65 rear-wheel horsepower at 5,000 rpm. Testing it at a friend's Porsche emporium proved all the more insulting. The mechanics laughed as the 400-cuber developed less power than the weakest under-3.0L Porsche engine they had ever run on the dyno. We smiled, knowing that the retest would be a different story.

Now Get to It

Start with the 400-inch engine. If that isn't feasible, get a 351M and locate a decent 400-inch crank and rods, which will drop right in. Either way, to displace 406 cubic inches you need the 4.00-inch stroke and the 4.030-inch pistons. Work with two-barrel heads exclusively. The ports in the four-barrel castings are far too large to run on the street—it would take a blower to fill them up and maintain port velocity. Along with their open-chamber design, the two-barrel heads—like those of a Chevy rat—have far larger ports than most small-block Chevrolets.

We selected high-quality, low-cost Silv-O-Lite cast pistons that were 0.030 inch oversize. The combination of the reverse-dome pistons, stock cylinder heads that were shaved 0.025 inch, and the 0.030 overbore produced a static compression ratio of 9.45:1. That's just about right for an iron-head street engine on pump gas. Our rings were nothing more than a $29 package with a moly-filled top ring, a cast second ring, and a three-piece oil-control set. That might have cost a little power at 8,500 rpm, but the slight drag at our theoretical 5,200-rpm peak wasn't worth the rapid wear rate of low-tension rings. This engine should stay together for at least 50,000 miles without your having to lay a finger on it.

The most critical area of the thumper is the camshaft. We needed one with about 220 degrees of duration at 0.050-inch lift, not too much overlap, and plenty of flow at low-to-medium valve lift. It had to open the valves quickly, raise them high, and then close them quickly without slamming the valve on the seat too hard. The Comp Cams X-Treme Energy #268 camshaft employs new reduced overlap technology, which is just the ticket for a smooth idle, and maximum power/torque between 2,000 and 5,200 rpm. (See specification comparison elsewhere in this article.)

The 268-degree camshaft fills the cylinders fast and lets them build pressure as if the engine had 11.5:1 compression. If passing the smog test is important, this type of camshaft will also be clean when it comes to the sniffer. The Comp Cams kit comes complete with a timing set, lifters, springs, retainers, locks, and the camshaft.

Sheldust Crankshaft & Engine in Pacheco, California, took the clunker apart and checked its dimensions. They also cleaned and inspected the block and heads prior to machine work. Then we got the damage report: The oil rings were locked solid in the piston grooves, five cam lobes were flat, and several guides were worn to death. Reasons enough why the engine mustered only 65 wounded ponies on the dyno. On the positive side, the block was perfectly sound and would clean up with the 0.030 overbore.

Sheldust measured the block from the center of the main bearing saddles to the outside corners of the deck and discovered a 0.015-inch difference between the largest and smallest measurements. That required decking both sides of the block at the head-mounting surface. The main saddles were straight, so the block wasn't align-bored or -honed, saving us $125. Sheldust bored the block 0.024 inch and finish-honed it to 0.030 inch using very fine stones to assure a good ring seal.

Next, we got into the act with a die grinder, abrasive rolls, and a couple of carbide burrs to match the gaskets to the intake and exhaust ports, as well as the intake manifold. We also cleaned up

At $125-$175, a standard three-angle valve job is money in the bank. If you lap the valves and assemble the heads yourself, you can save $50.

We used the intake and exhaust gaskets as porting templates for the cylinder heads and intake manifold. By doing the work ourselves, we saved time and several hundred dollars.

the combustion chambers and relieved a little of the rough casting behind the exhaust valve. While none of this delivers a dump-truck-load of torque, it does help promote a broader torque curve.

Instead of using tube headers (that would have been too easy), we spent more than 16 hours opening up the stock exhaust manifolds. If we were going to use headers, we would have selected a tri-Y design known for making torque.

Sheldust installed new cast-iron valve guides because they don't wear the valve stem like bronze silicon guides do. Moreover, cast-iron guides are best for the street with unleaded fuel and are far less costly. A three-angle valve job is simple to implement and yields excellent results. And at an average price of $125-$175, it's a good value. If you assemble the heads yourself, you save about $50. You can also lap in the valves by hand before assembly and get a better seat seal. We added stainless-steel valves to the recipe; they cost $150 but are lighter than the stockers and will hold a seal longer.

To clean it up, the crankshaft was turned 0.010/0.010 inch undersize, leaving the standard fillet radius on the sides of each journal. We polished the journals, and the grinder chamfered the oil holes at no extra charge. Considering its prime function, we decided not to balance the reciprocating assembly and saved another $175. We kept $150 more in our pocket by using press-fit wristpins rather than the full-floating type. Once the pins are pressed in (with the rod end red-hot), they stay in place and never work themselves into the cylinder walls.

The Melling oil pump was a bolt-on affair. The pickup was adjusted 3/8 inch from the bottom of a Canton 8-quart oil pan.

Puttin' It All Together

Assembly was straightforward. We installed the camshaft first to make sure it spun freely. (We prefer to turn the engine upright, with the camshaft bore facing the ceiling, to keep from scraping the cam bearings.) Our crankshaft was in spec, and after all the pistons and rods were installed, the assembly rotated with ease.

We used a die grinder, abrasive rolls, carbide burrs, and lots of sweat cleaning up the combustion chambers. The overall effect is a broader torque curve.

When we installed the cam timing gears, we checked the cam against the manufacturer specs. Most "production" shops no longer take the time to degree-in the camshaft, although we must note that the last 15 cams we've used have all been right on the money. Still, better safe than confused.

We used the stock rocker arms and Comp Cams pushrods during the build-up process. Make sure the pushrods turn by finger pressure when the valves are fully closed. If they don't, the machine shop hasn't measured the pushrods' stock height and subtracted the amount of material removed from the heads and the deck. Take this amount off the top of the valve stems. We could have bypassed this step with rocker arm studs and adjustable rocker arms, but they would have added $350 to our total cost.

We kept everything super-clean; followed the check-and-double-check procedures for the ring gaps, bearings,

Cost Review*

ITEM/SERVICE	PRICE
Holley carburetor	**$280.00**
MSD ignition and coil	**$175.00**
Intake manifold	$154.00
Cam and kit	$215.00
Stainless-steel valves	**$148.00**
Pistons	$98.00
Rings	$29.00
Clevite bearings (mains, rods, and camshaft)	$63.00
ARP rod bolts	**$35.00**
Fastener set	**$125.00**
Gasket set	$48.00
Oil pump	$25.00
Deep oil pan and pickup	$118.00
Abrasive rolls	$12.00
Core plugs (brass for water jackets)	$21.00
Bore and hone to 0.030-inch over	$155.00
Install cast-iron valve guides	$100.00
Three-angle valve job (seats and valves)	$145.00
Surface heads	$95.00
Install cam bearings	$47.00
Deck block	$55.00
Clean, inspect, and mag heads	$35.00
Clean and inspect block and rotating assembly	$85.00
Machine and polish crank to 0.010 inch/0.010 inch	$76.00
Press-fit pins	$60.00
Resize rods	$85.00
Our Total	**$2,484.00**
Maximum Savings	**$763.00**
Maximum Low-Buck Deal	**$1,721.00**

***Note:** Prices were established by shopping around, buying some parts from mail-order suppliers, and asking for discounts. Items in bold are not absolutely necessary for success, so their cost can be subtracted for maximum savings.

thrust surfaces, and other components; and then set the distributor on top dead center No. 1 before installing the intake manifold. But even with the long-block completed, we had lots to do before we could fire the engine. We needed fuel, and we needed fire.

We chose a high-performance mechanical fuel pump and plumbed in a good filter downstream of the carburetor. Using the formula for carburetor size—(engine cid [406] x max rpm [5,200]) divided by 3,456 (a constant)— we found that our engine needed 610 cfm. A 550-cfm Holley two-barrel delivers nearly that much air if you are restricted by configuration (as we were). A far better bet is a 650-cfm vacuum-secondary Holley spread-bore four-barrel, which gives excellent overall performance and good throttle response.

We could have kept the Ford ignition, but we used an MSD 6-AL ignition box and a Super Blaster Coil. The stock Ford electronic distributor works just great with the OEM ignition, and it passes smog on a street engine, but we splurged on the MSD goodies.

Since this engine motivates a '79 Ford Ranchero, we chose used dual catalytic converters (as per the factory) that flow 750 cfm each. A complete after-cat system by West Coast Muffler in Concord, California, gave us a 3-inch-diameter system joined with a two-into-one Flowmaster venturi Y-pipe in a single Flowmaster Big Block muffler. This system delivers more mid-range torque than a dual-exhaust system and costs less.

With a bag of torque, it isn't necessary to run a 5,000-rpm stall speed or 4.56:1 gears to get the car off the line. For $175, our skinflints installed an overused 3.25:1 limited-slip differential in the Ranchero's 9-inch housing. A key part of the overall combination is the transmission/torque converter package. We selected a 2,200-rpm stall converter to launch the car, but one with close internal clearances to maintain fuel efficiency during cruise mode. In high gear, this converter puts the engine at 2,280 rpm at 65 mph.

The bottom line? Poole's combination generated 375-plus lb-ft of torque at slightly more than 2,000 rpm—where most street driving is done. At 4,400 rpm, it peaked out at 456 lb-ft. In addition, the 406 generated 382 hp with all emissions gear intact. With a little fine-tuning, it would be easy to get 400 hp from this low-dollar thumper.

This time, we impressed the Porsche contingent. No Porsche has ever made 450 lb-ft of torque, or so much power, at such low engine speed. When we told them how much the project cost, they just about threw us out of the shop—$2,000 wouldn't even pay for a valve job on one of their client's dual-overhead-cam engines.

Even if you don't tread a righteous path, you can still make 400 hp and 450 lb-ft of torque with a street thumper. You can duplicate our engine combination for two grand and keep it "smog friendly" for everyday service. Just be sure to plan the complete package before you put hand to wrench. If you do, you can be rewarded with a super street thumper that takes no prisoners. **HR**

Street Engine Cam Specification Comparison

	COMP CAMS HIGH ENERGY 268-DEGREE CAM	COMP CAMS X-TREME ENERGY 268-DEGREE CAM
Part number	32-221-3	32-000-5
Grind number	FC 268H-10	FC 5433-5216 H110
Intake lobe	268 degrees/ 218 degrees at 0.050 inch	268 degrees/ 224 degrees at 0.050 inch
Exhaust lobe	268 degrees/ 224 degrees at 0.050 inch	268 degrees/ 230 degrees at 0.050 inch
Valve lift (1.73:1 rocker)	0.494-inch intake/ 0.494-inch exhaust	0.524-inch intake/ 0.529-inch exhaust
Lobe separation	110 degrees	110 degrees

Note: The Comp Cams High Energy 268-degree cam has been one of our favorite hydraulic grinds for years, and it's complimentary to the smog issue. The new X-Treme 268 is stronger, produces more power with less overlap, and has minimal pollution problems with the sniffer.

Continued from page 102

cally bolt-on horsepower from a 351-W or 351-C, you'll have to use aftermarket heads. The AR head combination for the 351-W is outlined above. The other option, for either a 351-W or a 351-C, is a set of SVO aluminum "high-port" heads, with a corresponding SVO 4-barrel intake (M-9424-A351 for the 351-W; M-9424-B351 for the 351-C). The heads come drilled for both Windsor and Cleveland water passages, with plugs to block those not used. On a production 351-Cleveland, the B351 manifold's valley cover must be trimmed straight at the back to match the block. The heads come in several port sizes to allow custom shaping, but the largest (M-6049-A3) can be used "as is" on a street engine.

Gary Hall of Hall Pantera has put together a 351-C package using these unported heads and manifold for street-driven Panteras which pulls 413 hp at 5200 rpm on Billy Williams' dyno, and 400 lbs.-ft. of torque from 2000 to 6000 rpm. With a flat-top piston (TRW L2379) and 62cc chambers, it's in the 9.5:1 range. Valves are Manley stainless, 2.19-inch intake (PN 11800) and 1.71-inch exhaust (PN 11805), springs are Isky PN 6005-A triples with the blue spring removed (125 pounds seat; 360 pounds open), and rockers are Norris stainless steel rollers (PN NPR-173) with SVO pushrods (M-6565-B342) and guide plates (M-6566-A302). Hall's HP-3 hydraulic cam (ground by Barnes) specs 208I/218E, .501-inch lift, and 110 lobe centers. For the induction, he bolts a Holley No. 3310 vacuum-secondary 750 to the SVO intake while the headers have 2-inch primaries and measure 29 inches long. The ignition uses a Ford D9TZ-1217-H distributor with 22 degrees mechanical advance (16 degrees initial), with a D5AZ-12A199-C electronic module. Dyno tests were run on 92-octane pump gas, through mufflers.

Depending on the heads, which swap, the basic recipes outlined here apply to either Clevelands or Windsors. If you follow them to the letter, we guarantee you'll get the same horsepower numbers—provided you know how to build an engine, tune a carb, set the cam and ignition timing, the weather is right, the stars are aligned, you're on a "good" dyno, and you're wearing the right color socks. Good luck! HR

SOURCES

AR Inc.
Dept. HR
P.O. Box 6358
Ventura, CA 93006
(805) 654-0654

B&A Ford Performance
Dept. HR
P.O. Box 6553
Ft. Smith, AR 72601
(918) 626-3997

Ford Motorsports Performance Equipment–SVO
Dept. HR
17000 Southfield Rd.
Allen Park, MI 48101
(313) 337-1356

Kaufmann Products, Inc.
Dept. HR
12418½ Benedict
Downey, CA 90242
(213) 803-5531

Hall Pantera
Dept. HR
15337 Garfield
Paramount, CA 90723
(213) 866-5852

Shooting for 300 Naturally Aspirated Rear-Wheel Horsepower

By James Lawrence

Bolt-On Dyno Thrash

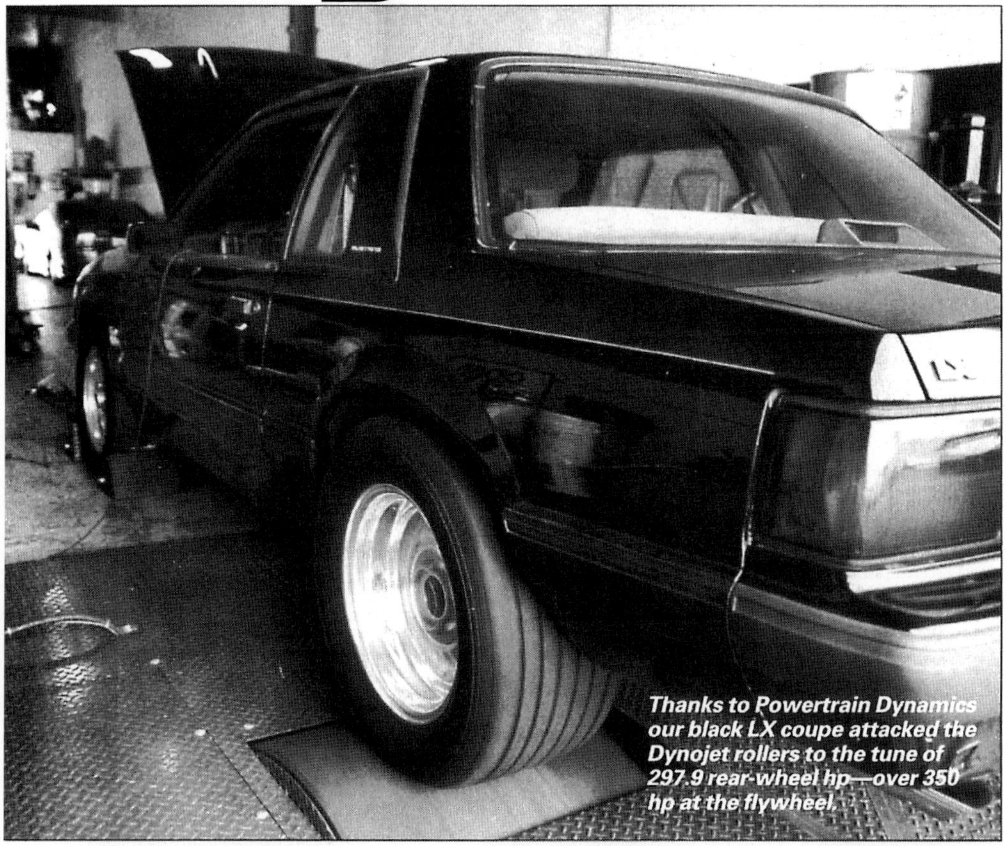

Thanks to Powertrain Dynamics our black LX coupe attacked the Dynojet rollers to the tune of 297.9 rear-wheel hp—over 350 hp at the flywheel.

In the old days, 300 naturally aspirated horsepower used to be the exclusive goal of the 5.0 elite. At the street races, or on cruise nights, any "all motor" Mustang that put up 300 big ones was rumored to be scary fast—and usually had a fairly lopey cam and couldn't come close to passing a smog check. Well, the times are changing. Cylinder head designs are improving, and if you want to be among the fastest stock-block-motor 5.0s at the street races, 300 horses just ain't gonna cut it, brother. If you're a street racer going for the funny money, plan on putting down 300 horsies at the *wheels*. We'll show you how to get there.

Bill Bassett, legendary mechanic at Powertrain Dynamics, in a deep moment of contemplation. The immense quantity of donuts and cheeseburgers consumed during the Bolt-On Dyno Thrash were not, contrary to popular opinion, enough to stunt Bassett's thought process.

Steve Ridout of Powertrain Dynamics handled all of the tuning duties for our '89 LX Coupe during the Bolt-On Dyno Thrash, as well as burning us a custom computer chip on the Dynojet.

The idea had been evolving at *5.0 Mustang* for a few months—show readers how to build a 300-rear-wheel-horsepower, naturally aspirated 5.0 Mustang around the stock short-block. With that much power, even at 3,200 pounds, you would be 11.90-capable. Instead of conducting a before-and-after build-up, we decided that it would be interesting to dyno test, step by step, each performance part that we added in our quest.

Our newest test mule—a five-speed '89 LX coupe—became the guinea pig. The car had some basic mods to start with—advanced timing, synthetic oil, 3.73 gears, Lakewood 90/10 drag struts and 50/50 drag shocks, removed front sway bar, Weld Draglight 15x3 and 15x8 wheels, and BFG 275/50-15 radials. On the Dynojet, the LX laid down a baseline of 196 rear-wheel horsepower and 269 lb-ft of torque. For each aftermarket part bolted onto the coupe, three separate dyno passes were run, and during each pass mild tuning (timing and fuel pressure adjustments) was performed to get the most power out of each component.

All the work, from strapping the LX to the Dynojet, to installing the parts, to tuning the combination was performed by Powertrain Dynamics, in Huntington Beach, California. We've worked with Powertrain before and were impressed by their attention to detail and expertise when it came to 5.0 Mustang performance. Bill Bassett, Powertrain's head mechanic, handled the installation work; David Lukason ran the Dynojet; and Steven Ridout was in charge of tuning the beast—the Powertrain crew did an admirable job helping our effort to build a 300-rwhp, all motor 5.0 Mustang!

BASELINE

191.6 hp, 261.9 lb-ft

Although we would have preferred to start with a 5.0 that was truly bone stock, rather than mostly stock, we got what we got. That was an LX coupe that already had some basic engine mods, like 16-degrees of timing, a K&N air filter, and Redline synthetic lubricants, no doubt responsible for our 5.0 exceeding the factory horsepower rating. We achieved 191.6 hp on Powertrain's Dynojet—about 10 hp more than an untouched 5.0 (stock EFI 5.0's usually spin about 175–185 horsepower).

MARCH UNDERDRIVE PULLEYS +8.6 hp, +11.1 lb-ft

200.2 hp, 273.00 tq

March underdrive pulleys replaced the stock pulleys on the water pump, alternator, and crank. A great bang-for-the-buck improvement, we picked up 8.6 hp and 11.1 lb-ft.

> **Performance: A** Report Card
> **Value: A**
>
> **What do they do?** Underdrive the engine accessories, such as the water pump, alternator, smog pump, and power steering pump—reducing parasitic drag and horsepower loss.
> **How long to install?**
> About 25 minutes.
> **Any special tools?** Air compressor/air ratchet to remove alternator pulley.
> **Comments/advice:** Great bolt-on addition with a quick install time.

Underdrive pulleys are often the butt of jokes among hardcore 5.0 racers, but we knew even before the first dyno pull that the March Street Pulley set was worth some power. We weren't surprised to see 8.6 horsepower and 11.1 lb-ft of torque improvement at the rear-wheels. We were very pleased with the fit and finish of the March pulleys.

Powertrain's Head Engineer Ridout told us that in his experience a 3-piece street pulley set is worth about 5–10 horsepower and won't cause any cooling or drivability changes. If you want more, you can install race pulleys, which can be worth a total of 10–15 horsepower—but can cause cooling and charging problems in street cars.

PAXTON FUEL PRESSURE REGULATOR +1.8 hp, +2.3 lb-ft

202.0 hp, 275.3 tq

> **Performance: B+** Report Card
> **Value: A**
>
> **What does it do?** Allows you to lean or richen the engine by adjusting the fuel pressure.
> **How long to install?**
> About 1.5 hours.
> **Any special tools?** No.
> **Comments/advice:** Adding an adjustable fuel pressure regulator is highly recommended but requires removing the upper intake manifold.

Stock 5.0 engines are shipped from Dearborn with overly safe and rich

Support Equipment

Any time you radically step up the power of your engine, you will quickly discover the weaknesses in your fuel system, cooling system, and drivetrain. In an effort to avoid delays and parts breakage, we planned ahead and took out some insurance—a fresh clutch, shifter, and fuel pump. Centerforce provided one of their 600hp-capable Dual Friction clutches, which use two different friction materials on the clutch disc and laugh at track abuse. BBK provided one of their in-tank 190-lph fuel pump kits to prevent engine starvation, and Steeda hooked us up with one of their awesome billet aluminum shifters to make sure we wouldn't miss gears during upcoming track testing.

Centerforce's Dual Friction clutch was another insurance item we installed. It will prevent any slippage or clutch problems under the brunt of more than 350 horsepower.

We installed Steeda's billet aluminum shifter to help us powershift Third without fail. We plan to track test our LX coupe when time permits.

Paxton's 5.0 EFI fuel pressure regulator (arrow) let Powertrain adjust our air/fuel ratio on the dyno to maximize horsepower during each of the dyno tests. Simply leaning out the stock motor made 1.8 hp and 2.3 lb-ft.

air/fuel ratios. Paxton's EFI fuel pressure regulator allowed us to adjust the fuel pressure from the stock 38 psi down to 36 psi. This produced a small horsepower increase, 1.8 hp and 2.3 lb-ft of torque. The adjustable regulator allowed us to tune in the proper air/fuel ratio for the further modifications.

Powertrain had to remove the upper intake manifold to install the Paxton unit—and it wasn't just a bolt-on. Due to the location of the vacuum fitting, we had to lengthen the vacuum hose, add a 90-degree hose end, and bend the fuel rail slightly for clearance. Paxton has since told us that newer designs eliminate this problem.

MAC 2½-INCH CAT-BACK EXHAUST +3.1 hp, +0.2 lb-ft

205.1 hp, 275.5 tq

Performance: B+ Report Card
Value: A–

What does it do? Increases exhaust flow over the restrictive and inefficient stock 2¼-inch mufflers, flow tubes, and tailpipes.
How long to install?
About 1.5 hours.
Any special tools? Sawzall (or hacksaw) to remove stock exhaust system.
Comments/advice: Impressive sound and performance. We recommend you install the high-flow H-pipe at the same time.

Powertrain's Dave Lukason bolts in the MAC 2½-inch Cat-Back system. The compact size of the MAC's Flow-Path mufflers is nice—but don't let their small size fool you, these mufflers made good power, adding another 3.1 ticks to our tally.

Eliminating the restrictive stock-cats should be one of your first modifications. Powertrain swapped this MAC hi-flow 2½-inch Catalytic H-pipe onto our LX coupe, and we were rewarded with 4.9 hp and 1.1 lb-ft.

We expected a small horsepower gain from the mandrel-bent MAC 2½-inch mufflers, considering the restrictive and inefficient stock 2¼-inch H-pipe. We ended up with an increase of 3.1 peak hp and 0.2 lb-ft of peak torque. It's worth mentioning that on the dyno the Motec data-logger indicated the free-flowing exhaust was making enough power to create a lean condition—we had to raise the fuel pressure to 42 psi.

The stock 2¼-inch H-pipe and four catalytic converters were enough restriction that the larger 2½-inch cat-back could be causing a loss in exhaust velocity—and prevented further horsepower gains. Think about buying and installing the entire MAC exhaust setup (cat-back and H-pipe) at once.

MAC 2½-INCH CATALYTIC H-PIPE +4.9 hp and +1.1 lb-ft

210.0 hp, 276.6

Performance: A– Report Card
Value: B

What does it do? Replaces the stock 2¼-inch H-pipe with a hi-flow mandrel-bent 2½-inch H-pipe.
How long to install? About 1 hour.
Any special tools? No.
Comments/advice: Expensive, but required to maximize our further modifications.

This was the final cork in the exhaust system, and getting rid of the cranky, kinked stock H-pipe was a blessing—giving us 4.9 hp and 1.1 lb-ft of torque at the peak on Powertrain's Dynojet.

The MAC 2½-inch cat pipe added a nice rumble to our exhaust, creating a much more throaty but still mild roar. Some of you might wonder why we didn't install an off-road H-pipe in search of impressive dyno figures? We felt that, with the year 2000 coming just over the hill, we'd better figure out how to make horsepower *and* make the smog monkeys happy. We'll eventually swap on an off-road pipe for a future story.

MAC 1⅝-INCH SHORTY HEADERS +5.2 hp, +8.3 lb-ft

215.2 hp, 284.9 tq

Performance: A Report Card
Value: A

What do they do? Eliminate the stock, crimped headers and substitute high-flowing, mandrel-bent 1⅝-inch unequal-length shorty headers.
How long to install?
About 2 hours.
Any special tools? No.
Comments/advice: A good bargain—mix of performance and value.

Stock 5.0 headers have a bad rap—assembled from crimped, cragly 1½-inch tubing. MAC's 1⅝-inch Equal-Length shorties on the other hand are a cut above—larger tubing, mandrel-bent, leading to a generously sized collector opening. Best of all, they carry an E.O. number that makes them 50-

Installing the MAC 1⅝-inch shorty headers took Powertrain about an hour, although MAC told us it takes the average enthusiast about two hours for the job. MAC's headers carry an E.O. number making them smog-compliant, and our '89 LX picked up 5.2 hp and 8.3 lb-ft at the wheels.

state smog legal.

We picked up 5.2 hp and 8.3 lb-ft of no-BS torque that will help you get down the track or roast your street skins—and for only 150 bucks we'll take that deal any time. Our dyno results showed that it's important to think of the exhaust system as a unit—improving one aspect of the system (headers, H-pipe, or mufflers) doesn't necessarily help when you have other areas being restricted. The power gain from the entire exhaust system was 13.2 hp and 9.6 lb-ft.

COMP CAMS BOLT-DOWN 1.7 ROCKERS +4.2 hp, +0.4 lb-ft

219.4 hp, 285.3 tq

Report Card

Performance: B+
Value: B

What do they do? Replace the flimsy stock stamped 1.6 rockers with extruded aluminum 1.7 rockers—increases valve lift.
How long to install? About 3 hours.
Any special tools? No.
Comments/advice: Installation is fairly complex and recommended for experienced enthusiasts.

There are three advantages when swapping 1.7 rockers onto a 5.0 engine: 1) They are lighter than the stock stamped steel jobs, helping increase horsepower and rpm capability, 2) the increase in rocker ratio, from 1.6 to 1.7, increases valve lift from .444 to .472, simulating a larger camshaft, and 3) the aluminum rockers are stronger and won't deflect under spring pressure.

We picked up 4.2 hp at the peak with the rocker swap, along with shifting the power band up. The Comp Cams rockers are adjustable via the pushrod cup. Though a nice feature, it also means that the rockers probably aren't great for the first-time enthusiast.

PRO-M 75MM MASS AIR METER, 24-LB-HR INJECTORS +3.8 hp, +1.7 lb-ft

223.2 hp, 287.0 tq

Report Card

Performance: B+
Value: B

What do they do? Open up the restrictive stock inlet track with a larger mass air meter and helps fuel the engine.
How long to install? About 2 hours.
Any special tools? No.
Comments/advice: Great modification for the beginner looking for a no-hassle, quick power increase.

On the Dynojet, the Competition Cams 1.7 rockers shifted the powerband upward, resulting in a 4.2 hp increase at the rear wheels.

As shipped from Dearborn, our LX coupe had some serious inlet side airflow restrictions—principal among those is the pithy 55mm mass air meter. We replaced it with Pro-M's 75mm mass air calibrated for the Motorsports 24-lb-hr injectors. Dyno results showed a 3.8hp and 1.7lb-ft increase at the wheels. No doubt the 60mm throttle-body and stock intake manifold are the next biggest airflow inhibitors.

EDELBROCK PERFORMER INTAKE MANIFOLD +15.3 hp and +9.5 lb-ft

238.5 hp, 296.5 tq

Report Card

Performance: A
Value: A

What does it do? Replaces the factory intake manifold with a completely redesigned, higher flowing cast aluminum upper and lower intake.
How long to install? About 3 hours.
Any special tools? No.
Comments/advice: Besides the underdrive pulleys, the best bang-for-the buck performance part we tested.

The Edelbrock Performer manifold is one of the most well-designed aftermarket products available for the 5.0 Mustang. It features a lower intake manifold with flow superior to the industry standard (both in total air flow and in port-to-port consistency), the GT-40 from SVO. Installing the intake was fairly easy, and the Edelbrock instructions were very detailed. After we fired up and ran through the gears a few times to warm up the engine, we

The Edelbrock Performer 5.0 intake manifold really boosted the power of our LX coupe—to the tune of 15.3 hp and 9.5 lb-ft, despite the restrictions of the factory cylinder heads, throttle body, and mass air sensor.

We decided to get serious and try Ford Motorsports' latest and greatest GT-40 aluminum "X" heads, with 58cc combustion chambers. Before installation on our LX coupe, we had the heads O-ringed, port-matched, bowl-blended, and converted to screw-in studs by Competition Heads.

In addition to port-matching the GT-40 "X" heads, we had Competition Heads convert the pedestal-mount heads to stud-mount, for use with Crane Gold 1.6 adjustable rockers.

We replaced the factory 60mm throttle-body and EGR spacer with a BBK 70mm unit, removing the last restriction in the factory air path. Our LX coupe's engine responded well with 10.2 hp and 3.8 lb-ft.

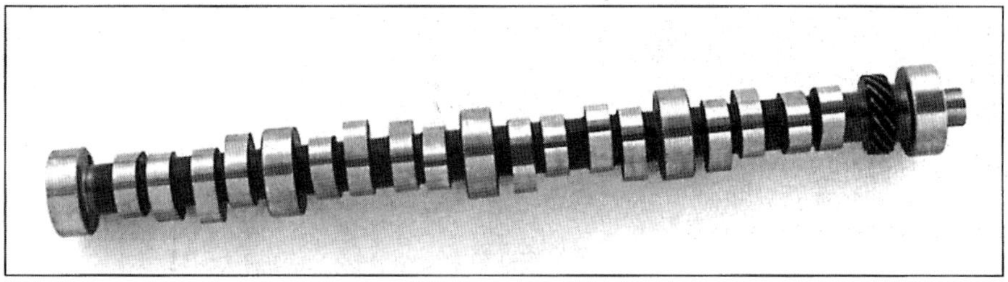

We used Motorsport's F303 camshaft, featuring .512 lift, 226 duration at .050, 288-degrees advertised duration, and a 114-degree lobe separation. It's got a fairly lopey idle, more so than an E303, and works great in our application, bringing the power band and shift point up to about 6,000 rpm.

were surprised at the amount of power increase.

Even breathing through the diminutive stock throttle body, the Dynojet showed an increase of 15.3 hp and 9.5 lb-ft of torque. Edelbrock designed the Performer intake for applications exactly like ours—we especially liked features such as the throttle-body opening being pre-sized at 70mm and the attractive new powder-coated upper manifold.

BBK 70MM THROTTLE-BODY AND EGR SPACER +10.2 hp and +3.8 lb-ft

248.7 hp, 300.3 tq

The next step in our Bolt-On Dyno Thrash was to replace the factory 60mm throttle-body, the major airflow restriction leading to the stock intake manifold. We used a BBK 70mm throttle-body and EGR spacers to pick up a surprising 10.2 hp and 3.8 lb-ft. Because we had already replaced the pint-sized factory mass air with a higher flowing 75mm Pro-M unit, the throttle-body had really become the cork in the system. This illustrates the value in paying attention to the entire airflow path in the 5.0, rather than changing one or two items.

Report Card

Performance: A
Value: A–

What do they do? Open up the restrictive stock inlet track with a larger throttle body.
How long to install? About 1 hour.
Any special tools? No.
Comments/advice: The intake tract, like the exhaust, is only as effective as it's weakest link. Think of the intake manifold, mass air, and throttle body as a system.

GT-40 ALUMINUM "X" HEADS, F303 CAM +41.1 hp and +23.9 lb-ft

289.8 hp, 324.2 tq

Probably the most serious modification you can do to a stock-block 5.0 is to swap on aftermarket aluminum heads and a hotter camshaft. For our Dyno Thrash, Powertrain selected a set of the new Ford Motorsport GT-40 "X" cylinder heads (58cc variety) and a Motorsport F303 camshaft. To help us reach our goal of 300 horsepower, we did some mild porting, as well as O-ringed the heads for Fel-Pro's Lock-Wire head gaskets.

We knew that the cylinder heads were going to be the key in our quest for the big 300 at the rear wheels. We selected the new GT-40 "X" heads for several reasons, but mostly because the 58cc combustion chambers will increase our static compres-

Report Card

Performance: A–
Value: B–

What do they do? Replace the stock heads and valves (1.78/1.45) with high-flow aluminum castings with redesigned ports, combustion chambers, and larger valves (1.94/1.54).
How long to install?
About 14 hours.
Any special tools? No.
Comments/advice: For the beginner or serious enthusiast—Motorsports' new GT-40 "X" heads turned out to be one of the best true bolt-on heads on the market.

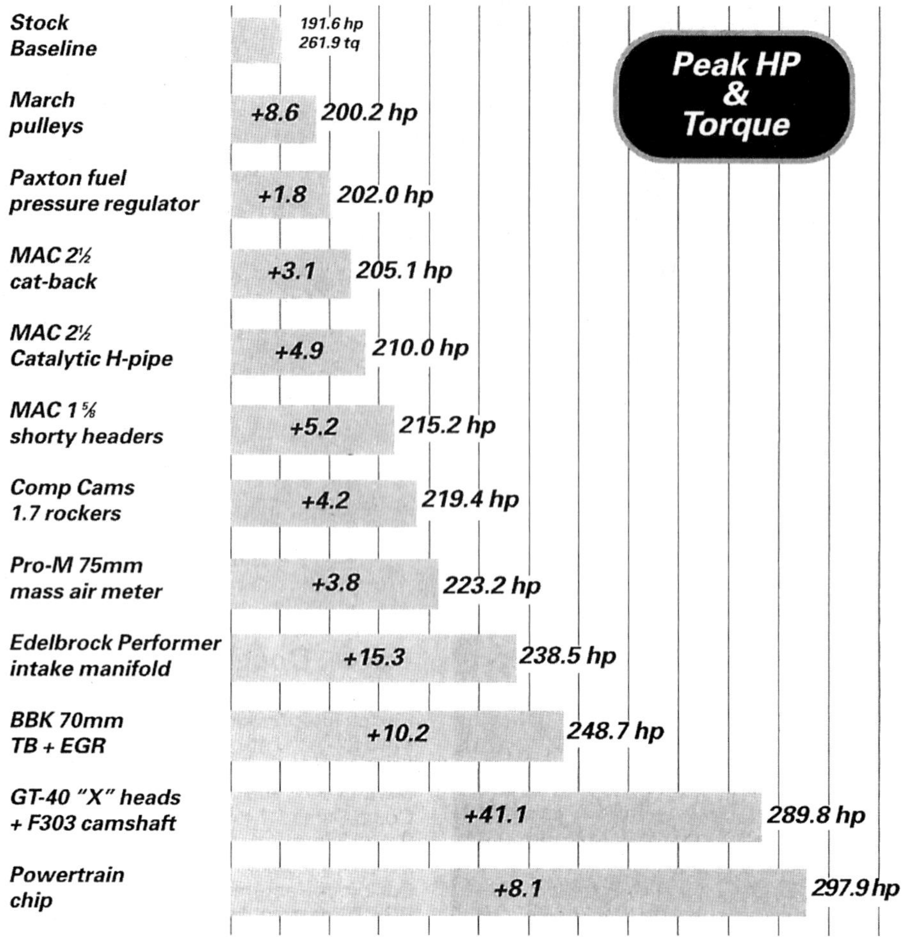
Peak HP & Torque
Stock Baseline
191.6 hp
261.9 tq
March pulleys
+8.6
200.2 hp
Paxton fuel pressure regulator
+1.8
202.0 hp
MAC 2½ cat-back
+3.1
205.1 hp
MAC 2½ Catalytic H-pipe
+4.9
210.0 hp
MAC 1⅝ shorty headers
+5.2
215.2 hp
Comp Cams 1.7 rockers
+4.2
219.4 hp
Pro-M 75mm mass air meter
+3.8
223.2 hp
Edelbrock Performer intake manifold
+15.3
238.5 hp
BBK 70mm TB + EGR
+10.2
248.7 hp
GT-40 "X" heads + F303 camshaft
+41.1
289.8 hp
Powertrain chip
+8.1
297.9 hp

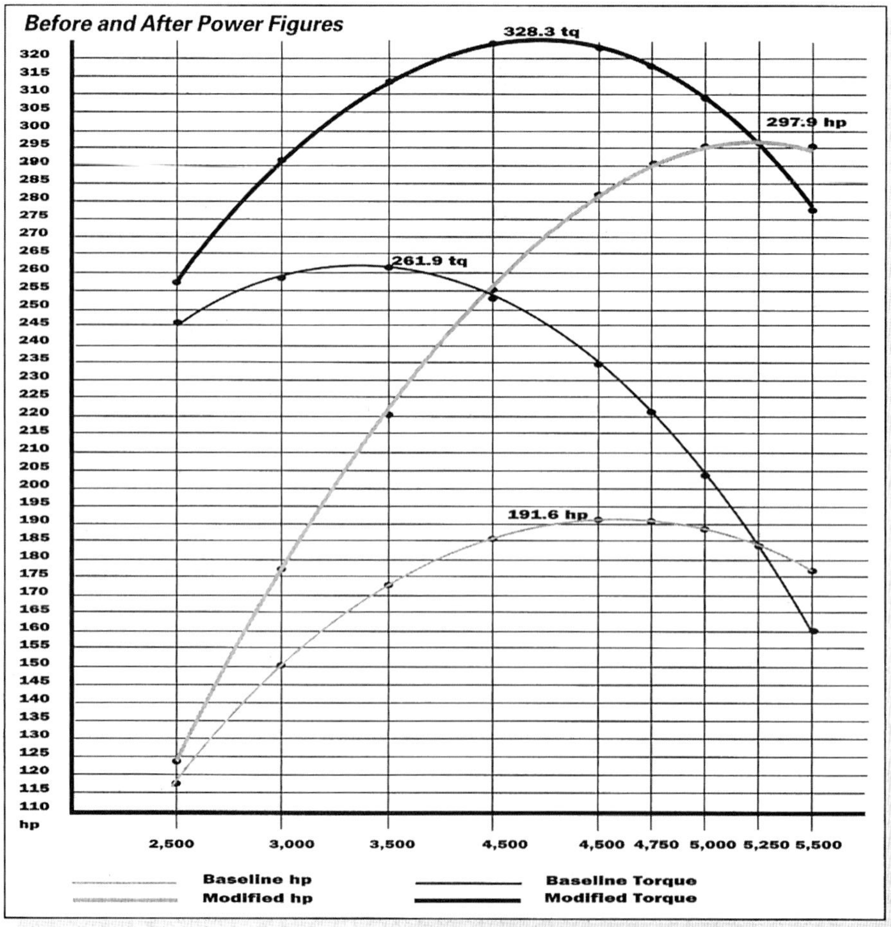
Before and After Power Figures
328.3 tq
297.9 hp
261.9 tq
191.6 hp
320
315
310
305
300
295
290
285
280
275
270
265
260
255
250
245
240
235
230
225
220
215
210
205
200
195
190
185
180
175
170
165
160
155
150
145
140
135
130
125
120
115
110
hp
2,500
3,000
3,500
4,500
4,500
4,750
5,000
5,250
5,500
Baseline hp
Modified hp
Baseline Torque
Modified Torque

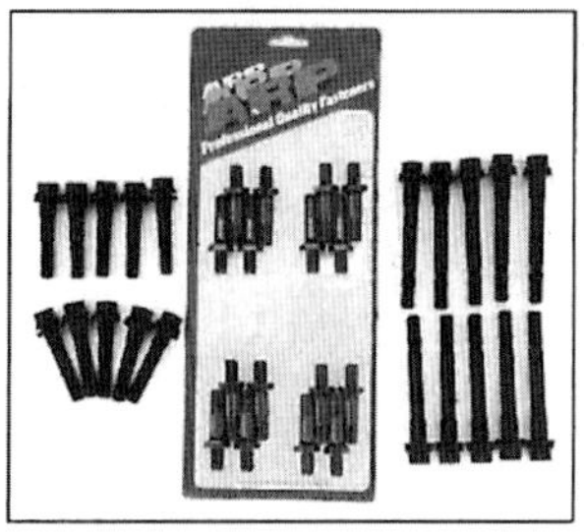

Powertrain recommends the use of ARP stepped-head head-bolts for the GT-40 "X" heads (PN 254-3708). They step from ½ inch to 7/16 inch, preventing the head from moving around on the gasket. We also used ARP 7/16-inch rocker studs.

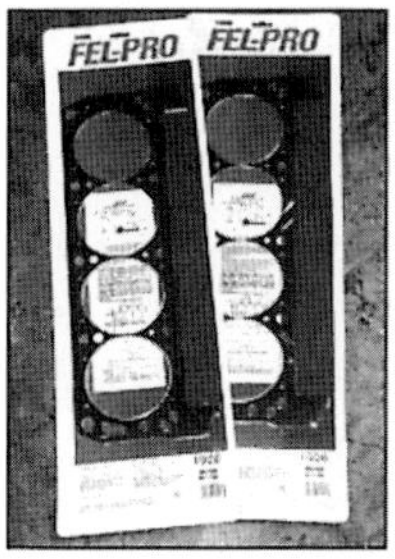

Fel-Pro No. 1006 Lock-Wire small-block Ford head gaskets were used for the head swap—along with a Fel-Pro No. 2804 performance gasket set.

sion to 9.5:1. In addition, the F303's 114-degree lobe separation will keep cylinder pressure elevated—and the mild porting and polishing will help ensure a high velocity air charge.

Bill Bassett did the head and cam swap in about a day, using Fel-Pro gaskets and ARP fasteners through the install. Powertrain's done this enough times that they were positive about the horsepower potential of our combination. We were all crossing our fingers as the LX was loaded onto the Dynojet. By far the biggest increase in power from any of the other bolt-ons, the GT-40 "X" heads and F303 cam made an additional 41.1 hp and 23.9 lb-ft of torque—giving us a total of 289.8 hp at the rear wheels.

POWERTRAIN 80MM MASS AIR METER, CUSTOM COMPUTER CHIP, 24-LB INJECTORS +8.1 hp and +4.1 lb-ft

297.9 hp, 328.3 tq

Sitting right at the 290hp mark (at the rear wheels) was pretty awesome, but we really wanted our LX coupe to hit that magic 300. Lukason suggested we try one of Powertrain's custom EEC computer chips, programmed using C&M Racing's chipmaster software, to recalibrate our engine control computer. Along with the chip, Powertrain installed their custom 80mm mass air sensor, which, because of it's precise mass-air voltage curve, allows them to achieve excellent idle characteristics

The Powertrain guys kept telling us that we could pick up some more horsepower and get closer to our goal of 300 rear-wheel horsepower with the combination of an 80mm mass air sensor (arrow) and a custom computer chip. This would allow Ridout to access the EEC-IV's fuel and timing tables.

Prices, Dough, $$$$$

March underdrive pulleys	$109.95
Paxton fuel pressure regulator	$119.95
MAC 2½-inch cat-back	$269.95
MAC 2½-inch cat H-pipe	$379.95
MAC 1⅝-inch shorty headers	$159.95
Comp Cams 1.7 rockers	$349.95*
Pro-M 75mm mass air sensor	$249.95
24-lb-hr injectors	$229.95
Edelbrock Performer intake	$495.95
BBK 70mm throttle body/EGR spacer	$249.95
GT-40 aluminum "X" heads	$1,250.00
Port-match, O-ring, stud-mount heads	$500.00
SVO F303 camshaft	$189.95
Crane Gold 1.6 stud-mount rockers	$279.95
Fel-Pro #1006 Lock-Wire gaskets	$119.95
Fel-Pro #2804 gasket set	$104.95
ARP #254-3708 head bolts	$114.95
ARP #134-7103 rocker studs	$37.95
Powertrain 80mm mass air meter	$249.95
Powertrain custom EEC-IV chip	$295.95

*Can be substituted for less-expensive bolt-down rockers.

All prices as sold by Powertrain Dynamics (714/373-0068).

along with impressive top-end power.

Unfortunately, the gods weren't willing—but we did pick up almost 10 hp to land us at 297.9 hp and 328.3 lb-ft at the rear wheels. So we came damn close to the magic 300, and we haven't even gotten a chance to try any tricks—like an off-road H-pipe, race pulleys, or running a short-belt. Watch for an update in a future issue, as we push to 300 and beyond.

Report Card

Performance: A
Value: B+

What do they do? Burn a custom computer chip that modifies the fuel and timing curves—also includes a larger mass-air sensor.
Any special tools? No.
Comments/advice: Changed our rough-idling beast into a smooth-rumbling pony.

Sources

Powertrain Dynamics
15628 Graham Street
Huntington Beach, CA 92649
714/373-0068

MAC Products
43214 Blackdeer Loop
Temecula, CA 92590
800/367-4486

Ford Motorsport SVO
44050 N. Groesbeck Highway
Clinton Township, MI 48036
810/468-1356

Edelbrock
2700 California Blvd.
Torrance, CA 90503
310/782-2900

Steeda Autosports
2241 Hammondville Rd.
Pompano Beach, FL 33069
954/960-0774

Centerforce
2266 Crosswind Dr.
Prescott, AZ 86301
520/771-8422

BBK Performance
1605 Railroad St.,
Corona, CA 91720
909/735-8880

March Performance
5820 Hix Road
Westland, MI 48185
313/729-9070

Paxton Automotive
1260 Calle Suerte
Camarillo, CA 93012
805/987-8660

Competition Cams
3406 Democrat Road
Memphis, TN 38118
800/999-0853

Pro-M/Best Products, Inc.
25760 John R. Road
Madison Heights, MI 48071
810/399-9223

Competition Heads
1004 Orangefair Lane
Anaheim, CA 92801
714/871-HEAD

MORE MUSCLE FOR FORD'S 351

THE SECRET TO BUILDING A GOOD STREET ENGINE IS COMPONENT COMPATIBILITY

HOT ROD BUDGET ENGINE BUILDUP

By Gray Baskerville

There isn't a hot rodder alive who hasn't been interested in bolting some real horseflash onto his otherwise dead player. "The trouble is," says Joe Sherman, low-buck engine builder extraordinary, "too many guys waste a lot of money buying parts they don't need. And there are plenty of shrewd countermen who will take advantage of their ignorance."

Edelbrock's Jim McFarland seconds the emotion. "Detroit designs their various powerplants around a certain mix of factory parts. These parts control induction, camshaft timing and exhaust gases. Unfortunate-

FORD'S FINE FAMILY OF 351S

Introduced in 1975 to simplify assembly-line procedures and parts interchangeability, Ford's 351M, or "Modified," motors are actually "high-block" or "raised-deck" family companions to the long-stroke corporate 400. Positive identification prior to starting work with any of these powerplants is absolutely necessary, because although many of an M's pieces will fit a 400, only a few interchange with Ford's identical-displacement Cleveland and Windsor motors. Both the 351M and the 400-incher can be visually identified by a ½-inch-wide by 1-inch-high vertical rib cast into the engine block at the immediate left (from the driver's seat) of the distributor. Experienced Ford floggers may also be able to spot the extra inch of material at the taller blocks' deck surfaces, from which the engine derives its high-block designation. Engines known to be factory-fitted to various FoMoCo car and truck models can be positively identified as to displacement by matching the fifth digit in the vehicle's identification or serial number with the corresponding number in that particular year and model's service manual. **HR**

HOT ROD BUDGET ENGINE BUILDUP

ly, the performance enthusiast who seeks to improve the output of his or her vehicle tends to be more interested in looking for big numbers than adding to the efficiency of these components."

The problem is simple. Few hot rodders are honest with themselves. Instead of taking the time to figure out what rpm range their engine usually operates in, or what kind of fuel economy they would like to achieve, or if it's torque rather than horsepower they want, they operate under the assumption that if they bolt on a set of headers, add an aftermarket intake manifold, attach a 4-barrel carburetor, recurve the distributor, and eventually stick in a high-performance camshaft, they'll end up with a significant power gain. That isn't necessarily true.

The fact of the matter is that most of us don't have the facilities to test these "efficiency mixes." We rely on word-of-mouth claims, or read about unbelievable performance figures. Most of these claims are pure bench racing; you know, wishful thinking. What we really need is some hard information, a form of "when the green flag drops, the bull stops." Consequently, HOT ROD has decided to take the bull by the horns—the first "bull" will be in the form of a late-model, 351-inch, closed-chamber Cleveland Ford (designated the "M" series), followed by similar articles dealing with a 350-inch Chevrolet V8 and a 360-inch MoPar in subsequent issues—and do one of our old "Prove it on the Dyno" bits. (Note: Some of the following modifications may not be legal for street-driven vehicles in your state.)

Now before we could do it on the dyno, we had to find a mule. That particular piece of horsepower was discovered loafing in HOT ROD Publisher Holly Hedrich's '77 Ford Ranchero. However, the two-year-old V8 was more than tired and had to be completely rebuilt and blueprinted before it could be lowered onto Edelbrock's hook. According to Kurt Hooker, Edelbrock's dyno technician and the guy who rebuilt Holly's hound, "It wasn't in too good a shape. We hot-tanked the block, straightened the bores with a Speed-Pro-recommended No. 825-grit Sunnen hone (used in conjunction with SP's moly ring combination), and took .005-inch off the deck. While the main bearings were no problem, the rod bearings were

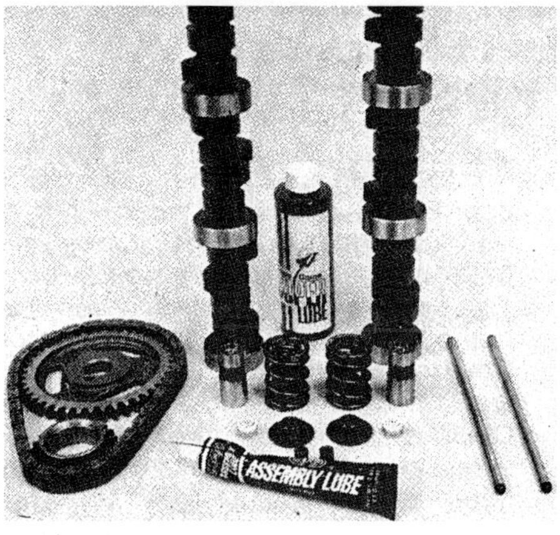

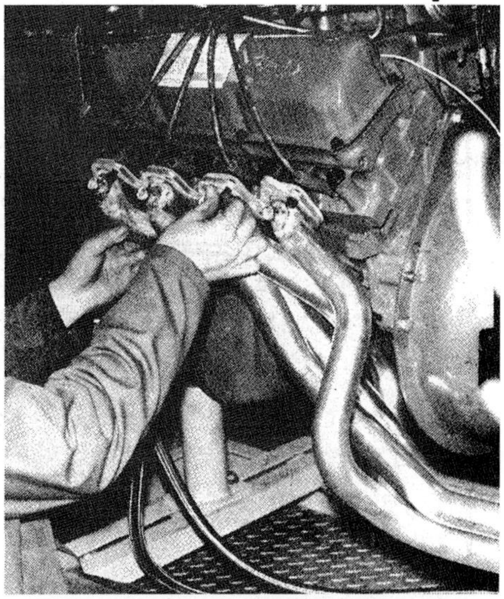

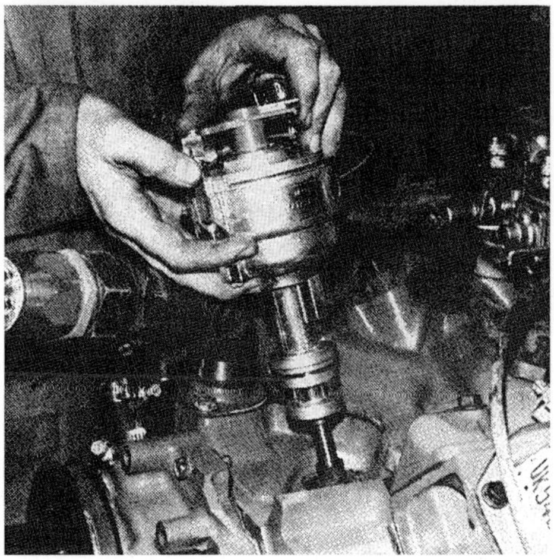

Stock 351M block was decked (.005-inch), had bores straightened and honed; the lifter bores were cleaned up; and it was painted with Rustoleum damp-proof primer.
Crane supplied high-rpm Fireball 302A and HT-288-2-NC mid-range cams.

After break-in procedure that involved varying throttle settings for three hours, Kurt Hooker installed a set of Champion RBL 17Ys. Accel igniter was given a total of 24 degrees advance at 3800 rpm. Hooker headers made a big improvement in torque, picking up 14 lbs.-ft. at 2750 over stocker. A 600-cfm Holley carburetor was used in conjunction with both Edelbrock manifolds.

MORE MUSCLE FOR FORD'S 351

shot, because there was too much crush. We also cleaned up the pistons and Magnafluxed all the moving parts, then shot-peened and polished the rod beams, and miked the crank for straightness and roundness.

"Once we dropped in the crank and checked the bearing clearance (.002-inch), I covered the end seals with RTV Silicone to keep them from possibly leaking oil, then torqued down the main caps to 105 ft.-lbs. I also checked the end play (.004-inch) and connecting rod clearance (.015-inch) prior to torquing the rod caps to 45 ft.-lbs. By the way, the ⅜-inch nuts were all cracked and failed the 'Mag,' so we replaced them with new ones. As for the cylinder heads, we simply cleaned up the ports, gave them a good valve job, checked the guides, made sure we had 90 pounds of seat pressure, and torqued them down (in an inside-out sequence) to 105 ft.-lbs. The lubrication is stock; we torqued the stock intake to 25 ft.-lbs."

What "Hook" didn't tell you was that he augered the crank dampener onto the snout with an 8-pound hammer.

In essence, the plan was to design an induction, cam and exhaust system to produce a significant increase in the engine's torque range that one would encounter under normal driving conditions. To determine what would be best, we thought we would home in on both the engine's torque curve and its Brake Specific Fuel Curve, to measure its combustion efficiency. Torque was considered important, because that's what you feel on the seat of your pants. The Brake Specific Fuel Curve (BSFC) is determined by dividing uncorrected horsepower into how many pounds of fuel are being consumed, then finding out the ratio of burned fuel to power output. For example, a .453 figure means .453-pound of fuel burned per horsepower at a given rpm. In a nutshell, all the BSFC tells you is how much fuel the engine is using to produce maximum torque. If an engine uses 100 pounds of fuel per hour to produce 1000 ft.-lbs. of torque, then it's more efficient than an engine that uses 150 pounds of fuel to produce the same amount of torque.

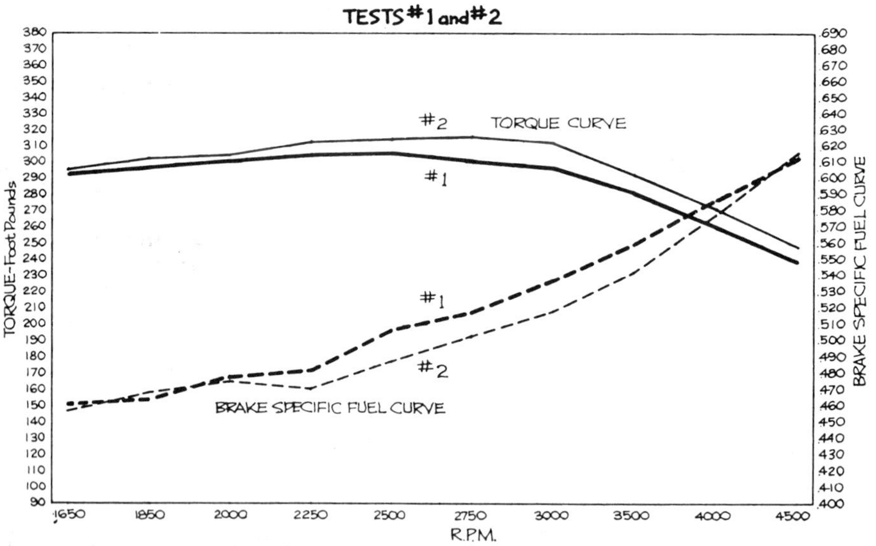

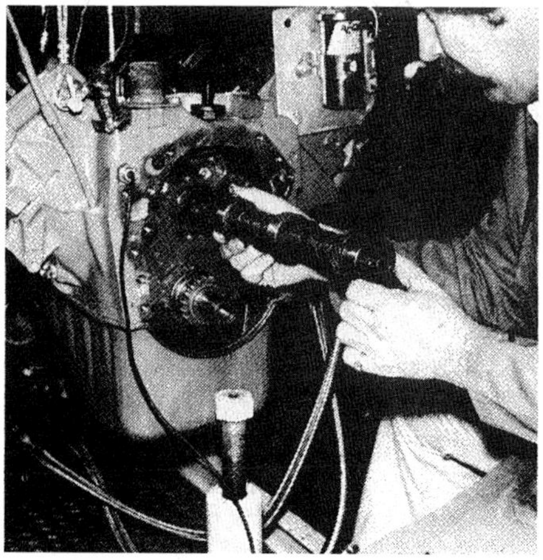

Henry Walther of Crane Cams installs the Fireball 302A. Cam timing killed bottom-end torque, but allowed the 351M to churn out 302 horses at 5000 rpm.

"Hook" gets ready to slap on the Edelbrock SP2P intake manifold. Combination with stock cam put out 333 lbs.-ft. of torque @ 2750—a jump of 27 over OEM 2-barrel system.

Our game plan was almost as simple. We would start by breaking in and baselining the 351 Ford to determine how fuel-efficient the V8 was in its stock configuration, then proceed from there in typical hot-rodding fashion; that is, add a set of exhaust headers, stick in a recurved distributor, switch to an aftermarket intake manifold supporting a 4-barrel carburetor, and finally, make a cam change. In addition to these bolt-ons or bolt-ins, we thought it would be interesting to put together two basic packages. The first would be our high-rpm combination that would feature a set of Hooker headers, one of Accel's distributors, an Edelbrock Streetmaster intake manifold/600-cfm Holley 4160 carburetor (part No. 8006) 4-barrel induction system, and a Crane Fireball 302A hydraulic cam. The second combination would be more conservative and would employ the same headers, with an additional 3-inch extension of the collectors to improve low-end torque, the same distributor, an Edelbrock SP2P intake manifold/8006 Holley combo, and a low- to mid-range Crane HT-288-2-NC hydraulic cam.

TEST NO. 1: After a three-hour break-in period, the baseline run was made, using the stock 2-barrel carburetor. As you can see by the accompanying graph, the 351's maximum torque (306 lbs.-ft. @ 2500 rpm) featured a BSFC of .506. For you horsepower freaks, this run netted 207 horsepower at 4500 rpm.

TEST NO. 2: The only addition to the otherwise stock Ford M was bolting on a set of Hooker No. 6125 headers. These pipes feature 1¾-inch-by-31-inch-long tubes that dump into a 3x10-inch-long collector. Now look at the torque jump all the way through the entire rpm range. Obviously, a reduction in back pressure resulted in a maximum torque figure of 316 lbs.-ft. @ 2750, while the BSFC reads .502 , only a .004-pound decrease over the stocker at 2500 rpm. Horsepower was 214 @ rpm.

TEST NO. 3: The next step

was to replace the stock electronic ignition with an Accel No. 34202 distributor recurved to give 24 degrees of advance at 3800 rpm, or a total of 36 degrees (with 12 degrees in the crank) overall. We also twisted in a set of standard-gap Champion RBL 17Y plugs gapped at .032-inch. Stockers were Champion RBL 17Y6 gapped at .060-inch (for the electronic ignition). Again the 351M liked the change. Maximum torque was 327 lbs.-ft. @ 2750, and the BSFC dropped to .487 at the same rpm. Horsepower was 215 @ 4500 rpm.

TESTS #3 and #4

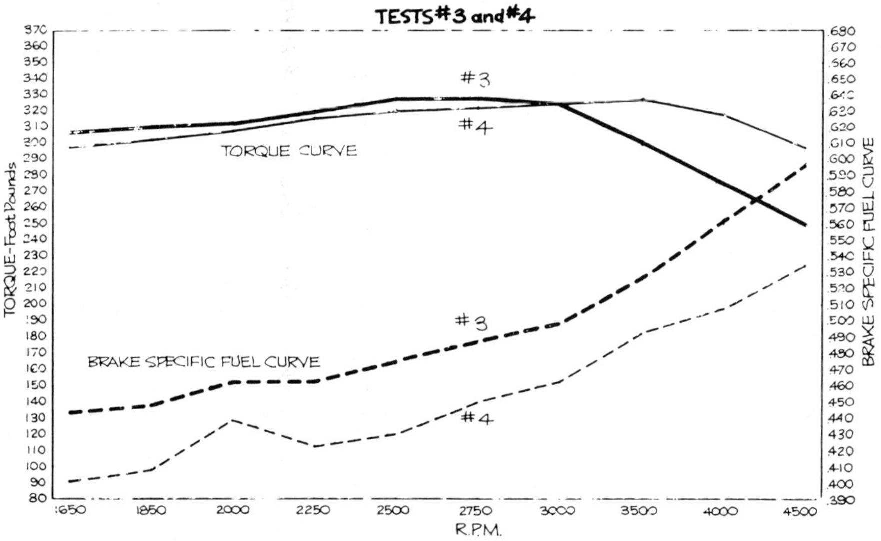

TESTS #5 and #6

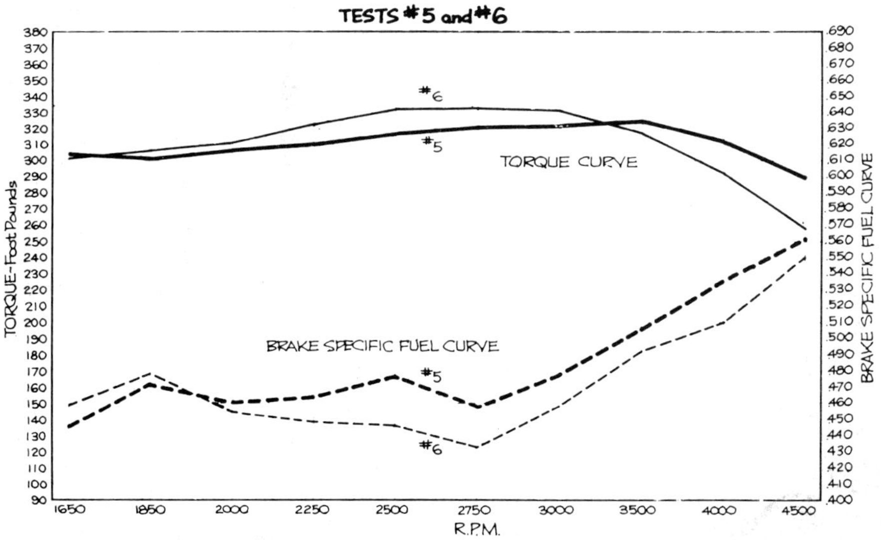

TESTS #7, #8 and #9

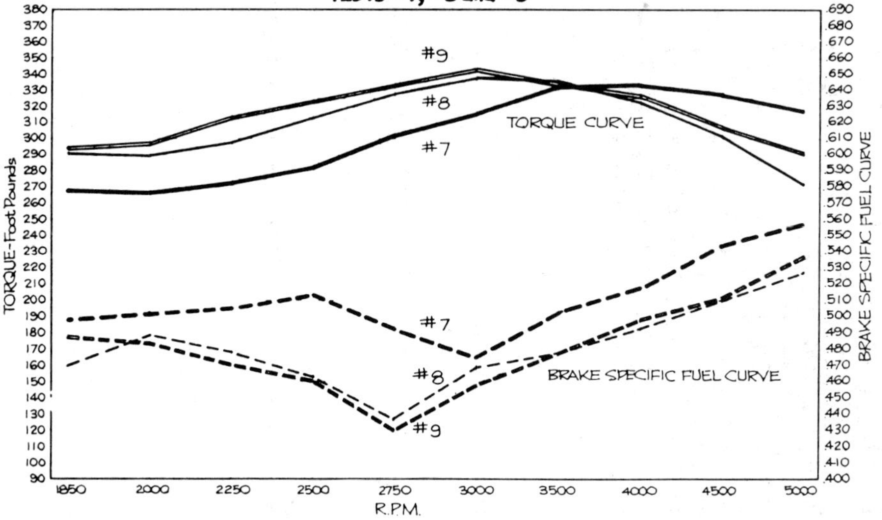

TEST NO. 4: Now we added a Streetmaster fitted with a 600-cfm Holley carburetor. The only changes made to the carburetor were to richen the primaries to No. 65 Holley jets and to change the secondary metering plate to a No. 12, .003-inch richer than stock. However, the overall torque curve suffered considerably, and we moved the curve up to 3500 rpm. Then the corrected horsepower figure climbed from 215, with the stock induction system, to 256 @ 4500 rpm. BSFC was .493 at maximum torque.

TEST NO. 5: Because of the increased flow characteristic of the Streetmaster, the carburetor was further altered to provide more gas; the primaries were richened up an additional two jet sizes (No. 67), but the secondary metering plate (No. 12) was retained. Torque dropped off 4 lbs.-ft. @ 3500 rpm, and the BSFC went up to .505 at the same rpm. Maximum horsepower at 4500 was 251. The feeling here was that while the runner moved more air, the engine's volumetric efficiency was lessened. In fact, it would produce a "hole." What was need was more cam timing.

TEST NO. 6: We now switched to Edelbrock's new SP2P intake manifold (designed to work with stock cam timing and jetting, and featuring a 4000-rpm ceiling), and the carburetor was returned to its box-stock setting (No. 62 jets primary, No. 39 secondary metering plate). Maximum torque jumped up to 333 lbs.-ft. @ 2750, while the BSFC dropped down to .431 at the same rpm. Horsepower was 222 @ 4500 rpm. This was clearly a significant change in engine efficiency.

TEST NO. 7: This was the test we all had looked forward to. In essence, this was to be our high-horsepower run, and it featured a combination that included Crane's Fireball 302A hydraulic cam. According to Crane, this particular grind exhibits a fair idle, with good low- to mid-range torque and horsepower. The recommended usage of this cam is that it be combined with a small 4-barrel carburetor, an aftermarket intake manifold, recurved ignition and headers. Lift is .505 on the intake, .516 on the exhaust, with 224 degrees on the intake, 232 degrees on the exhaust. The cam has 108-degree lobe centers. The other change was that we went back to the Streetmaster, using the same carbu-

Continued on page 125

SOUTHERN FRIED FORD

HOW TO BUILD A 570-HP 351C FOR A SUNDAY DRIVE

By Mario Rossi

Editor's Note:

There are few forms of auto racing that impose more demands on an engine builder than NASCAR Grand National competition. Not only is it essential for these engines to produce enough power to be competitive; they must also make this power in a specific usable rpm range that is sufficiently wide to accelerate those massive stockers out of the corners and reach peak speeds down the straightaways. But perhaps of even greater importance is the demand on these engines to live under sustained high-output usage—to go 500 miles flat out.

There's quite a difference between building an engine to produce peak power for a matter of seconds, as in drag racing, and one that must perform flawlessly for three and a half hours. Consequently, the talents and knowledge of those men who have accomplished such feats are greatly respected. One such man, Mario Rossi, currently builds engines at the spacious Reed Cams, Inc., facility in Decatur, Georgia. Because Mario's accomplishments include extensive work with the Ford 351 Cleveland engine, we've asked him to outline what is required to turn one of these engines into a NASCAR screamer. His approach is methodical, leaving nothing to chance. It is a procedure that could be applied equally to any make of engine, and therefore, what he has to say is of value to anyone building engines, Ford or not.

I don't build street engines and I don't build drag engines. Consequently, a lot of what I go through to build an engine may seem totally unnecessary to many of you, but believe me, it's attention to little details that usually makes the difference between being in the running at the end of a 500-mile race or pushing the car behind the wall at some point before that. You simply can't take chances or compromise where durability is at stake. What I'll be describing here is a 355-cubic-inch, single four-barrel, NASCAR Grand National engine based on the Ford 351 Cleveland. It will have a bore of 4.020 inches, a 3.50-inch stroke, a billet crank, a mushroom flat-tappet cam and 12.5:1 compression. An engine built *exactly* as follows, using these exact components and procedures, without shortcuts, will produce a reliable 560-to-570 hp at 7000-to-7200 rpm with 445-to-450 ft.-lbs. of torque at 5600-to-5800 rpm. This is equivalent to what the engines in the top-running cars are developing.

The basic foundation of this, or any, engine is the block. Until recently the only suitable block was the 351C-4V block, which has four-bolt main caps. Unfortunately, cylinder wall thickness, block rigidity and main webs thickness were all marginal on these blocks. Additionally, casting quality, in terms of core shift and flashing, along with machining tolerances, left a lot to be desired. Now there's a better block available. It is generally referred to as the Australian block, but that's kind of a misnomer. My understanding is that the block is cast in the U.S., shipped to Australia for machining, and then shipped back to the U.S. This block has beefier main webs and caps. The caps are high-nodular iron, the cylinder walls are thicker (.165-inch minimum) and more uniform, and the pan rails are thicker for greater rigidity. Core shift seems to be virtually nil, and the castings are very clean, with little flashing. They also seemed to be machined to very close tolerances. These blocks bear a casting number, XE-182540, and they are currently available from either Gapp & Roush Performance in Livonia, Michigan, or Hank the Crank in North Hollywood, California.

The top block is a production 351C-4V version, whereas the bottom one is the Australian block. Compare the main webs and the pan rail in terms of material thickness.

These Australian blocks are fairly expensive, and due to the extra material, they are considerably heavier than the conventional 351C blocks, but the advantages so outweigh these drawbacks, both in durability and the mandatory preparation time, that in my opinion, there is absolutely no reason to even consider using the conventional block. However, even as good as it is, I still check the Australian block meticulously before beginning any preparation work.

I begin by removing all the plugs in the block (oil passage and freeze plugs). The cam bearings, which come with the bare block, are also removed. I examine the block closely for any production flaws, core sand in the water jackets, etc. The oil passages are inspected for machining chips or restrictions. They are also checked to verify that all passages are drilled to the correct diameter and depth.

Next, the main bearing saddles are checked for proper alignment. Then the caps are torqued to the correct specification and the main bearing bores are checked for diameter and concentricity. Each lifter bore is checked for diameter and to verify that its centerline is perpendicular to

PHOTOGRAPHY: C.J. BAKER

PART 1

the centerline of the camshaft. The cylinder bores are similarly checked to verify that they are perpendicular to the crankshaft centerline. This is crucial. If the bore is canted slightly to the front or rear, the piston pin will inevitably hammer out the pin locks and destroy the engine. If they cant up or down, the piston will tend to cock in the bore, cracking the cylinder wall, causing bearing wear, etc. In the event that you do find a slight misalignment and the cylinder bores are to be overbored, such as I do by .020-inch, then it *may* be possible to machine out that misalignment, using a boring machine that locates off the crankshaft centerline. Other items checked for perpendicularity are the cam thrust surface and the bellhousing mating face. They must be per-

Before installing the bearing inserts, the edges are chamfered to clear the crank journal radii and the oil slot is enlarged to correspond to the oil passage indicated. Don't overdo it. Note that restrictors have also been placed in the passages supplying the cam bearings to limit oil flow in that area.

The billet crank, steel rods, pistons and pins are all from Hank the Crank. Note the use of heavy metal to achieve proper balance while keeping the counterweights smooth and streamlined.

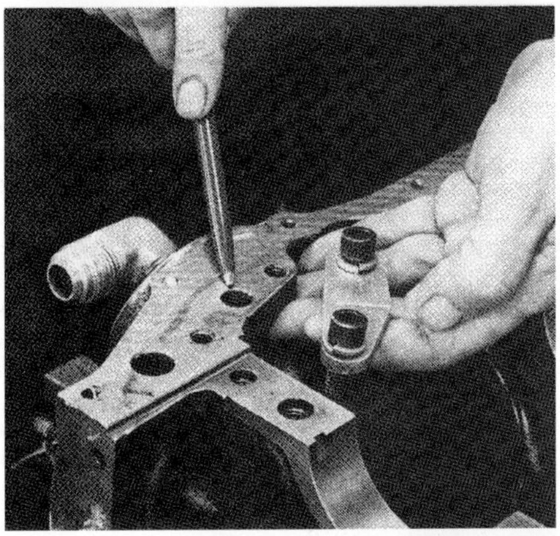

When modifying the block to accept a dry-sump oiling system with remote oil filters, it is necessary to plug or cap this oil drainback hole to prevent oil from leaking out of the engine. Rossi generally uses a small aluminum plate, as shown, to accomplish this. Although not shown, the hole in the front of the block for the dipstick tube is also plugged when a dry sump is used.

The Ford 302 harmonic balancer (part No. C8FE-6316-B) is used, but the stock counterweight (left) is machined away, since it is not needed with the fully counterweighted crankshaft.

All honing is done in a Sunnen CK-10 using 1½-inch head plates torqued to 110 ft.-lbs. Sunnen JHU-820 stones (400 grit) are used with a honing load of approximately 30 on the machine's meter to produce the final wall finish of 8 to 13 microinches without burnishing.

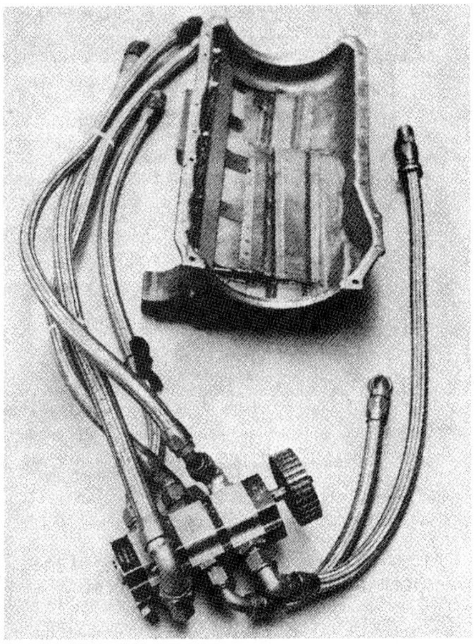

A Weaver three-stage dry-sump oil pump with 1.7-inch gears is used with a special Aviad oil pan. Note the extensive use of screens and scrapers in shallow oil pan.

pendicular to the camshaft centerline and the crankshaft centerline, respectively.

All of this checking is very time-consuming, but it is essential in terms of durability. Additionally, you are determining if you have a good block to work with. If you find a problem, then you have the option of fixing it or getting another block before proceeding. In checking many of the Australian blocks, I find that most have turned out to be "dead on" perfect, but occasionally you will find a small problem, so check it carefully. Now we're ready to begin the deburring and machining operations, along with the correction of any problems.

First I grind off all casting flash and sharp edges. (I don't like to cut my hands while working on engines.) Any rough edges or stress risers, particularly in the main web area, are carefully radiused or smoothed to reduce the likelihood of cracks. All threads are chased with a tap, including the head bolt threads, which are also checked for depth to verify that the head bolts will not bottom. If the block requires align-honing of the main bearing bores, that is done next. Then the bottoms of the lifter bores are spotfaced to accept the mushroom tappets.

When you are spotfacing the lifter bores, they must be cut sufficiently deep to allow .030- to .040-inch clearance between the bottom of the lifter and the nose of the cam at peak lift to allow for overrev protection. Another quick way to get adequate clearance is to cut them deep enough so that a straightedge laid on the cam bearing saddles (without the bearings in place) will clear the tappets.

Next the block is bored and honed for a finished size that will give a .009- to .0085-inch piston-to-wall clearance. (I'll cover the pistons in detail later on, but now I'll detail the boring/honing procedure.) All boring and honing is done with 1½-inch headplates installed, using the exact type of head gasket that will be run on the engine and torquing the bolts to 110 ft.-lbs. The main caps are also installed and torqued. I prefer to use studs in place of the stock capscrews, but the capscrews are acceptable. The block is bored to within .010-inch of the finished size. It is then set up in a Sunnen CK-10 and honed to within .0015- to .002-inch of the finish size, using the Sunnen JHU-625 or JHU-5525 stones. It is then finished to size with the Sunnen JHU-820 stones, maintaining a stone load of 25 to 30 on the machine's honing load meter. This produces a very smooth finish of 8-to-13 microinches without burnishing.

The next major block machining operation is the squaring and decking of the block, but to do this you'll have to have the crank, rods, pistons, pins and bearings, so I'll cover those before getting into decking.

Actually, the crank, rods and pistons are some of the simplest parts of the engine from the standpoint of preparation, since they are all supplied by Hank the Crank. I use the billet, fully counterweighted, Tufftrided crank. Hank also has a fully reworked cast crank available for less money, but even with all of the tricks, these cast cranks are good for no more than one superspeedway race. On the other hand, the billet cranks will last five or six races (some have even been run 8 to 10 races), so in the long run the billet is cheaper and more reliable. The billet crank has large radii, which means that the

LUBING THE CLEVELAND

The design of the lubrication system in the 351 Cleveland engine seems to have its priorities reversed, supplying oil first to the lifters and cam bearings, and then delivering whatever oil volume and pressure is left to the main bearings, crank and rod bearings. This provides a lot of opportunity for both pressure and volume to bleed off before reaching the critical bearings. To offset the problem, many engine builders resort to running excessively high oil pressure in the Cleveland (over 100 psi) to ensure adequate oil to the crank. But there are better ways to do it.

One of the better cures for the Cleveland seems to be the installation of an external oiling system, like the one offered by Hank the Crank. This system, in conjunction with a series of metering and restrictor plugs, delivers full oil volume and pressure to the rear of the block, where it is manifolded directly to the main bearings through a set of pressed-in sleeves in the lifter galley. The system works with either a wet- or dry-sump oil system.

An alternative is to restrict the oil going to the lifters with the conventional oiling system by sleeving the lifter bores. Hank also offers a kit to do this. Either way, it is much-needed help for this engine.

Restrictor plugs are inserted to limit oil to the cam bearings and left lifter bank.

Although it is an earlier design, plumbing in this kit indicates how oil is routed.

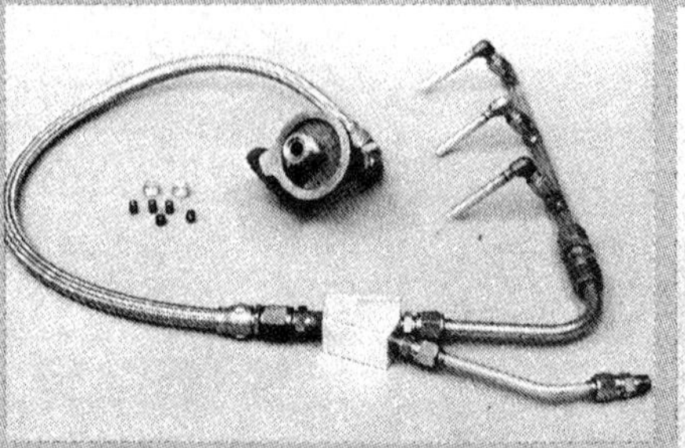

These are the basic components of the external oiling kit in its latest form.

If the external oiling kit isn't used, the lifter bores can be sleeved to limit oil.

bearings must be chamfered. The TRW/Clevite 77 fully grooved inserts are what I prefer to use, since quality control and sizing seem to be very good. The oil slot in the upper shell is also opened slightly, as shown, to improve oil flow. About the only thing to check once you receive the HTC crank is straightness, since it may have gotten bent in shipment. If it is within .001-inch, it is okay.

Hank also supplies the steel rods, which are of his own design. If the engine is being built for either Daytona or Talladega, where a narrower rpm range can be utilized, 6.0-inch rods are used. For all other tracks, the rod length is 5.80 inches. These rods are bushed on the small end. They come completely finished, including SPS bolts, but check 'em for length, as well as straightness and concentricity.

The pistons are a BRC/HTC design. They feature a high-silicone-content aluminum alloy, a 10cc dome and a wide skirt, and the top ring groove is .425-inch down from the deck to allow adequate material above the groove and the intake valve pocket to prevent failure (.100-inch after cutting for valve relief). The pistons are available for either the 6.0- or 5.8-inch rods, and they already have all of the necessary pin oiling holes. HTC lightweight taper pins are used and are retained with double .043-inch Spiroloks.

With the crank, rods and pistons on hand, a trial fit can be made to determine block deck height. This is done by installing the crank and fitting the main bearings. Using bore gauges and micrometers, get a true main bearing clearance of .0025-to .00275-inch, and then mark the inserts so that you can put them back with the right cap and journal on final assembly. Fit the pistons to the rods with a pin clearance of .0008-inch in both the rod and piston. The end clearance on the pins should be zero. After running, this pin end-play will open up to .003- to .005-inch, which is correct. Next, fit the pistons in the bores, using plenty of oil, but no rings. Set the rod bearing clearance at .0025-inch, again using TRW/Clevite 77 bearing inserts. At this point, the piston deck height can be checked to determine how much the block will have to be machined. Check rod side clearance at this time too. It should be .017- to .020-inch. If it isn't, get it. Then disassemble everything and set it aside.

The deck height that we will want is .010-inch down in the bores. It is also important that the deck be cut exactly perpendicular to the cylinder centerlines. This is usually accomplished by taking a light cut on the block to square it, and then a second cut is taken to achieve the correct deck height. Most decking equipment leaves a suitable surface finish for good gasket sealing.

The final steps in the block preparation include a light chamfer at both the top and bottom of the cylinder bore. It will also be necessary to install oil restrictor plugs (see sidebar).

This completes the machine work on the block, but we're still a long way from final assembly. The piston domes still have to be smoothed, valve reliefs must be checked, and if necessary deepened, and the rotating assembly has to be balanced. We'll cover all of that, along with the heads, valvetrain, induction system, etc., next month. In the meantime, there are a few items that are still required for the lower end. I use the HTC external oiling system, since it bypasses all of the shortcomings of the stock Ford oiling system (see sidebar). However, the oil pump used is a Weaver dry-sump unit with two scavenge and one pressure stage. Each stage uses 1.7-inch gears. A few tips on this system include a 1-inch-i.d. line from the oil reservoir to the pressure stage and dual ¾-inch-i.d. lines from the scavenge pumps to the tank. Use only straight or 45-degree fittings in all of the oil lines. If a 90-degree fitting must be used, select a free-flowing aircraft fitting. The pump is driven with Gilmer belt, using a 26-tooth gear on the crank and a 32-tooth gear on the pump.

Ground clearance problems prevent the use of an ideal pan, in terms of depth, but the pan used does have a full-length screened windage tray and scrapers. Pickups are provided for three scavenging stages, although only two are used. This Aviad pan is available from Reed Cams, Inc.

When you've completed all of the required machining, do a super-thorough cleaning job on the block. This will include using bottle and rifle brushes in the oil and water passages, along with liberal use of solvent and detergents. If a steam cleaner is available, use it, particularly on the cylinder bores. When a block is properly cleaned, you should be able to dry it off and wipe a clean white handkerchief in the cylinder bores without *any* trace of gray appearing on the cloth.

I'm sure this is enough to keep you busy until next month, so start gathering the pieces. It all takes a lot of time. **HR**

SOUTHERN FRIED FORD

CONCLUDING THE IN-DEPTH DISCUSSION OF HOW TO BUILD A GRAND NATIONAL-TYPE 351C FORD

By Mario Rossi

Last month we managed to cover the majority of our 355-inch 351C Ford short-block, but looking back, I feel that I should mention a few points. First off, when fitting the crankshaft in the block to check clearances, make sure you have .004- to .006-inch of crankshaft end play. Next, the flywheel that I recommend with this combination is a 22-pound lightweight steel version (part No. GR7W-6380-A) from Gapp and Roush. And last, I recommend Magnafluxing or some other inspection process. Magnafluxing, X-ray and Zy-Glo checking is good insurance, and absolutely essential on a rebuild or overhaul.

In this business of racing, there's nothing quite as constant as change. We're always looking at new things and exploring new ways of achieving greater power levels and durability. Since Part 1, we've done considerable dyno work and found something of interest. If you recall, I specified using 6-inch rods only for high-speed tracks like Daytona and Talladega. However, in light of increased torque and horsepower figures we've been able to achieve through a broader rpm range, I now recommend a 6-inch rod for all applications except very short ½-mile tracks, where the 5.8-inch rod should be used. Not only does this combination produce more power, but the longer-length rod improves piston and cylinder wall durability. The difference isn't big, but if you've got the choice, go the longer route.

When we left off last month, we'd done everything we could without the cam and heads, but to proceed, those pieces will have to be obtained next. The cylinder heads are another thing that take little preparation on the part of the engine builder. They are supplied, completely ready to assemble, by Reed Cams, Decatur, Georgia. These heads are based on 6F18 castings and are fully ported. They flow 330 to 340 cfm on the intakes, 215 to 220 cfm on the exhausts, measured at .600-inch valve lift at a depression of 28 inches of water on a Superflow bench. The ports are not welded, although the heat riser opening is welded closed. These heads are supplied with 2.19-inch-diameter Ford/TRW titanium intake valves with 11/32-inch stems and Manley 1.71-inch (11/32-inch stems) stainless steel exhausts. The valves are cut with a three-angle valve job, and cast-iron guides are installed with .002-inch clearance on the intakes and .0025-inch on the exhausts. Perfect Circle rubber valve seals with Teflon inserts control valve guide oiling.

The combustion chambers on these heads are held between 64 and 65cc, which when used in conjunction with the .010-inch down deck height, a piston dome of 8 to 10cc and a .035-inch-thick head gasket, will yield a compression ratio of 12.2:1 to 12.5:1. For additional strength, the Ford heads also must have the combustion chambers "pinned" (see sidebar) to prevent cracking during endurance racing. Reed Cams does this as part of their normal head preparation. To complete the heads, Reed steel spring cups, titanium retainers, hardened steel locks and valve springs are used. The valve springs, which are Reed No. 2260 outers and No. 2261 inners with a

PHOTOGRAPHY: C.J. BAKER

Reed Cams modified the heads, welding the heat riser closed and fully porting and polishing them for maximum flow. The heads are assembled with titanium intake and stainless steel exhaust valves, Reed springs and components, and roller rockers.

PART 2

flat damper, have 145 pounds seat pressure and 360 to 380 pounds open pressure, depending on the maximum valve lift. They are installed at a height of 1.87 inches.

The head gaskets used on this engine, McCord No. 6850M versions, have a compressed thickness of .035-inch. This gasket is stamped to indicate which end must go toward the front of the engine. It is possible to install them backward, so be careful or the water jacket openings will be misaligned. I also drill a 3/16-inch hole in the gasket corresponding to the water jacket (as shown) at the front of the engine, to promote more even cooling. On final assembly, I also coat the block, head and both sides of the gasket with high-temperature aluminum paint to ensure a good seal.

The next area of concern is the cam and valvetrain. Once again, these pieces come from Reed Cams. We are presently using cams with intake and exhaust durations of 274° and 280° (at .050-inch lift) for Daytona and Talladega and 268° and 276° for the short- and intermediate-length tracks. These cams are ground on 106° lobe centers, with lifts of .675-inch for the intake and .685-inch on the exhaust (long duration) and .657-inch and .676-inch for the short-track grind. Both are mushroom tappet cams. Of course cam design and technology are constantly evolving, so by the time you read this, even better profiles may be available, so check with Reed for the best cam for your particular application. Since gear drives aren't NASCAR-legal, I use a Cloyes True-Roller chain and sprocket, but if you can legally use a gear drive, do so.

Because of the mushroom tappet cams, limiting cam "walk" (end play) is critical to prevent the tappets from overlapping onto adjacent cam lobes. This is done by machining either the cam sprocket or the thrust plate. I try to hold the end play to .002-inch, but up to .010-inch is acceptable.

The remainder of the valvetrain consists of the Reed mushroom tappets (supplied with the cam), with a 1-inch face and an .872-inch body. The Reed ⅜-inch-diameter 4130 steel alloy pushrods, Reed 7/16-inch, heat-treated 4330 steel alloy rocker

The Reed mushroom tappet cam provides flow and minimal duration for peak power across a wide rpm range. For a detailed explanation on how mushroom tappets work, see "Magic Mushrooms," elsewhere in this issue.

THE RIGHT TRACK

Another important aspect of cylinder head preparation has to do with the type of track the heads will be used on. For example, when the heads are ported and flowed, flow is checked at valve lifts of .300-, .400-, .500- and .600-inch and a flow curve is plotted. If the heads are prepared to produce optimum flow at .600-inch lift, then the flow at the lower lifts tends to be somewhat less than if the heads are ported to achieve the highest possible average flow at all valve lifts. The difference is that a port that flows exceedingly well at .600-inch lift usually has to have a bigger cross-sectional area. Consequently, the velocity of flow is slower at the lower lifts, thus making the low lift flow sort of lazy. Therefore, heads for superspeedways, such as Daytona or Talladega, where a very narrow rpm range is utilized, are ported solely for peak flow at the maximum valve lift checking specification of .600-inch. Heads for the shorter tracks, where a wider rpm range is needed, along with quick throttle response and torque coming off the corners, are ported for the best average flow, even though peak flow at .600-inch may be off 10 or 20 cfm. ●

KEEPING A HEAD

Preparing the Ford 351C cylinder heads for endurance racing requires a bit of extra effort. To withstand 3½ hours of continuous racing, the combustion chambers and surrounding gasket surface need to be beefed-up to prevent cracks and blown gaskets. Such reinforcement is called pinning, and it consists of eight support studs screwed into the head until they bear against the backside of the chamber and gasket surface. The accompanying photos indicate where they are placed. One stud is inserted at the end of the head, passing through the water jacket adjacent to the intake valve. Three other studs are installed by drilling and tapping through three of the four exhaust rocker arm stud bosses. The fourth stud boss is not used because of the heat riser passage directly beneath it. The four remaining studs are installed right through the gasket surface of the head, passing through the water jacket and bottoming against the exhaust port floors. The head is then surfaced to cut these four studs perfectly level with the rest of the gasket surface. ●

Eight pins are installed in each head. One pin is screwed through the end of the head (left), while four others are inserted directly through the gasket surface (circles) to bear against the bottom of the exhaust ports to provide additional support to the gasket surface, which is milled flat after the pinning is completed. The remaining three pins are inserted through three of the intake rocker stud bosses until they bear against the gasket surface at the points indicated (Xs). Note that no pin is installed at the middle of the head because of the blocked heat riser in that location.

studs, Harland Sharp 1.73:1 aluminum roller rocker arms and a Reed rocker stud girdle (features adjusting nuts with positive locking setscrews) complete the assembly, along with stock Ford pushrod guide plates.

With all of these pieces on hand, it is time for another trial assembly to check proper valvetrain geometry and piston-to-valve clearance. Install the cam bearings, the mushroom tappets for one cylinder, the cam and cam drive (degreeing the cam at the straight-up, on-time position), the crank and one piston/rod assembly without rings. Bolt on the head for the cylinder being checked and install the remaining valvetrain pieces for that cylinder, setting the valve lash to normal running clearances. Using checking springs, measure the piston-to-valve clearance at the closest point, which usually occurs 8 to 12° BTDC for the exhaust and 8 to 12° ATDC for the intake valve (on the overlap stroke). With today's state-of-the-art mushroom cam designs, with their inherently quick "off-the-seat" action, it is almost always necessary to remachine the valve pockets in the pistons to get a minimum of .080-inch and .100-inch clearance on the intake and exhaust valves, respectively. If you don't take the time to check this and to get the proper clearance, don't complain when you stuff the valves into the piston domes and junk the engine. If you plan on swapping back and forth between cam designs, set the piston-to-valve clearance with the wildest grind, which will ensure adequate clearance for milder cams.

After cutting the additional valve relief depths, the tops of the pistons should be deburred and radiused to remove any sharp edges and to promote low-lift flow and flame propagation within the combustion chambers. This is also a good time to get the rings and check them for land clearance and end gap. I use the Sealed Power/Speed Pro rings, since they have shown consistently good results in this type of application. The top ring is the BR-18PF-12 +.025, 1/16-inch ductile iron, plasma-sprayed, moly-filled, conventional-twist, barrel-faced version—end-gapped at .014-inch. End gaps are measured ½-inch down the bore, with head plates torqued to spec. The oil ring is the SS-50U-567 +.025 low-tension 3/16-inch version. Side clearance on both the top and second moly rings should be .001-inch; the ring groove depth should be such that the face of the ring can be pressed back between zero and .004-inch below the ring land. Under no circumstances should the ring extend beyond the ring land when fully bottomed.

With all the basic engine components gathered, fitted and machined, have the rotating and reciprocating assemblies balanced to 50 percent. I don't subscribe to overbalancing for engines of this type. The final step is to make sure everything is meticulously clean and begin final assembly. Remember that when using a mushroom tappet cam, the first thing that must be installed in the block are the mushroom tappets. Use standard Ford crankshaft end seals and be sure to install the main and rod bearings on the same journals they were fitted to earlier. The SPS capscrews in the connecting rods should be torqued 90 to 95 ft.-lbs., and a good lubricant should be used on the threads and under the capscrew head. I use a mixture of light oil and molybdenum disulfide. If you've got the equipment to measure it, rod bolt (capscrew) stretch is a better indicator of proper tightening than torque. The proper stretch is .004- to .0045-inch. I also use the moly/oil mixture on all bearings and rubbing surfaces during assembly.

Initially install the cam at ½-degree advanced and then recheck it after the heads are in place with all of the valvetrain installed and the valves lashed to the correct specification. The strain placed on the timing chain by the valve springs normally retards the cam about ½-degree, thus putting the cam straight up, where it belongs. If that's not where it is under these conditions, get it there.

Another tip on assembly is to pin the freeze plugs so that they can't pop out. The Waterman Racing Engines aluminum freeze plugs, which incorporate an O-ring, along with a small retaining bar that positively bolts the plug into place, are an alternative.

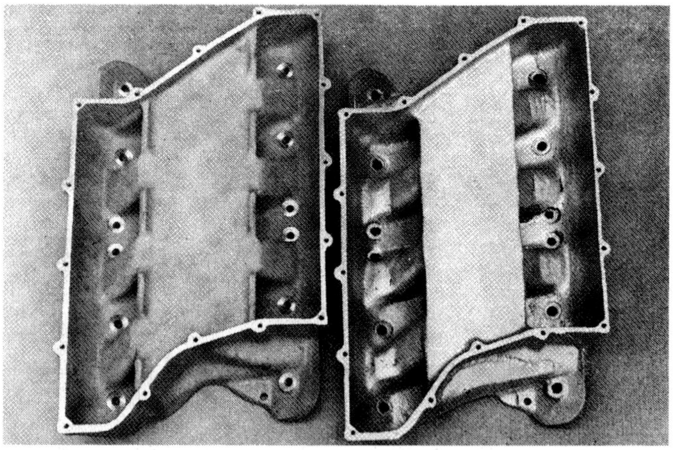

The Bud Moore box manifold is modified (right) to smooth and turn the flow into the runners, which are balanced and matched as much as possible. Notice how the bottom runner wall, in particular, has been moved out.

Before and after comparison of the piston after decking, valve pocket cutting, smoothing and polishing.

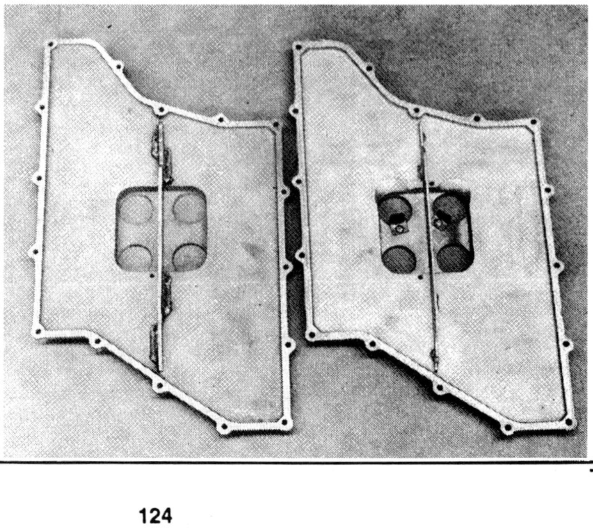

The manifold lid is also modified with an extended divider, flow tabs and directional radii adjacent to two of the four openings. Extensive flow bench is required to achieve peak flow and equal distribution on hese manifolds.

SOUTHERN FRIED FORD

All that's left now is the final dress-up equipment. The first item to consider is the intake manifold. To date, the Bud Moore box manifold has shown the best power and torque for the 351C in this type of racing. The box design yields itself to the wider bore spacing of the 4500 Dominator series of carburetors.

Reed does a lot of work to the Bud Moore manifold to improve flow and distribution. Ridges are removed, radii ground and runners reshaped. To do it properly, the distributor even has to be reworked to clear the modified manifold. Other tests have shown that more power can probably be obtained using a properly reworked 4150-series carb, if a good manifold can be developed to work with the standard-flange Holley carb. The actual carburetor I use is supplied by Reed Cams. This 4500 has the mandatory 1 11/16-inch restrictor rings, but it is completely flowed and matched on Reed's flow bench. The standard Ford intake gasket set that uses the one-piece oil shield/gasket is used in bolting it all together.

The heads are topped off with either stamped steel or cast aluminum valve covers. I prefer the Ridgeway ribbed cast aluminum covers, since they are sufficiently rigid to support the necessary breather tubes, quick-disconnect oil fitting and dry-sump vent that must be used.

The water pump is modified by cutting down the o.d. of the impellers

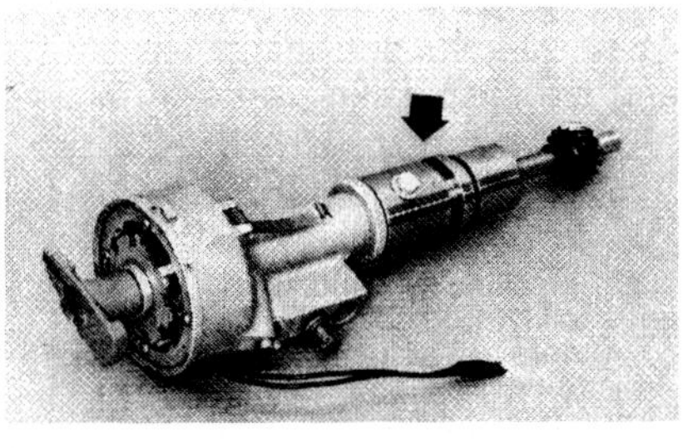

The distributor shaft and housing must be extended to clear the modified manifold.

about ½-inch. This is assuming the use of the Boss 302 cast-iron impeller. This reduces cavitation and slows the water circulation for better cooling at high engine speeds. It also releases an extra 5 horsepower at racing speeds. But the fuel pump is something else altogether. The stock Ford mechanical fuel pump has a durability problem, but we're restricted to mechanical pumps. Consequently, Bud Moore has contracted with Carter to distribute specially made Carter mechanical pumps. This pump is a must, but be sure to take your checkbook when buying one. If you can legally run an electric fuel pump, position one at the rear of the car too.

The rules also require that a production type of ignition be used. If you can find one, use the Ford magnetic impulse transistorized unit, which has a tach drive distributor. These units are rare and expensive, but they work well. The plug wire is silicone-jacketed 8mm by Jack Cotton. Champion BL-57 plugs are selected for the superspeedways, and BL-60 for short tracks. In both cases, they are gapped at .025-inch. As with the fuel pump situation, if you can legally run an aftermarket electronic ignition, do it.

The final item to consider is headers. I use 2-inch-diameter primaries, 34 inches long, feeding into a 3½-inch collector. The tailpipe is also 3½ inches in diameter, usually between 40 and 50 inches long. However, 16 to 18 inches behind the primaries I install a 3-inch-diameter "H" pipe between the collectors. This fools the exhaust system, creating a situation similar to the collectors opening to atmosphere at that point instead of into the tailpipes. This is good for 8 to 10 ft.-lbs. of torque in the mid-range.

That about wraps up our southern-fried 351C Ford. It's an expensive engine to build, but any Grand National engine is. Properly done, it's a durable and powerful engine, capable of winning races in the right chassis with the right driver. And you just can't ask for more than that. ●

A 3/16-inch gasket hole here improves cooling. The dipstick hole is plugged too.

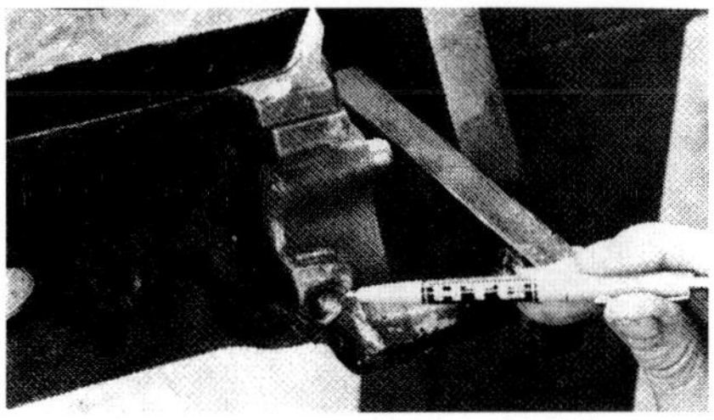

The Australian block doesn't have this hole for the clutch linkage. Be sure to drill and tap it before installing the engine.

MORE MUSCLE FOR FORD'S 351

Continued from page 117

retor jetting as in test No. 5. However, the torque characteristics of the engine changed significantly. Maximum torque of 333 lbs.-ft. @ 4000 rpm moved the peak output well up the rpm range, and the BSFC moved up accordingly to .518. However, the engine output climbed to 282 horsepower @ 4500 rpm (a jump of 31 horsepower over the stock cam/Streetmaster combo) and continued to a maximum of 302 @ 5000. We could have buzzed it up a little tighter, but "Hook" chickened out, and we were never able to see what the ol' Henry could really do.

TEST NO. 8: This run featured two basic changes to the 351, in the hope of coming up with a good low- to mid-range combination. First we installed Crane's HT-288-2-NC hydraulic cam (encompassing 288 degrees duration on the intakes, 298 degrees duration on the exhausts, .496-inch lift on the intakes, .517-inch lift on the exhausts with 112-degree lobe centers) that was said to provide good, low-end torque. The second change was to reinstall the SP2P intake/Holley 600 with stock jetting. As you can see, the swap proved bountiful. The torque curve picked up considerably, to 3500, and provided a maximum of 337 lbs.-ft. @ 3000 rpm and a BSFC of .460 at the same rpm range. Horsepower was 258 @ 5000 rpm.

TEST NO. 9: The final test was made to see if extending the header pipes and collector an additional 3 inches would result in an increase of torque. At 3000, we picked up 6 more lbs.-ft (343), while the BSFC stood at .457. Obviously, the change in length improved the torque in the 2700-3000-rpm range while dropping the BSFC. Horsepower at 4500 rpm was 263.

Obviously, we could have played with various combinations until we wore out either Holly's hound or my typewriter. However, Kurt Hooker wanted to know what combination Holly wanted before they returned the 351 to the Ranchero. Jim McFarland's answer was significant: "Let Holly look at these dyno figures; then let him decide!" **HR**

FORD POWER

WEIAND'S NEW FORD INTAKE/CAMSHAFT COMBOS ARE SUPER STRONG

By John Baechtel

Dyno testing performed by Air Flow Research established the strong performance potential of Weiand's new small-block Ford "Action Series" manifold/camshaft combinations.

TEST 1

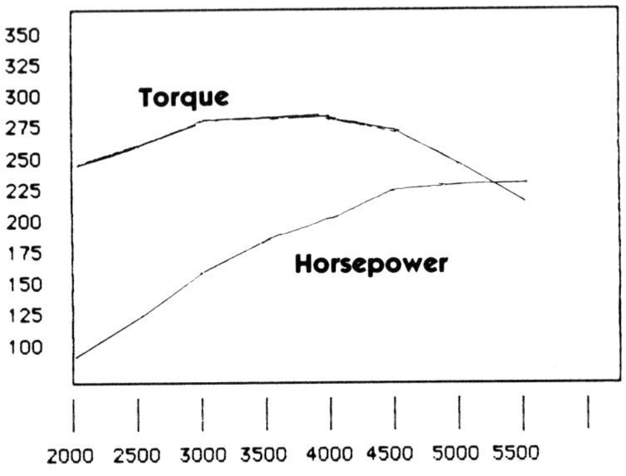

The Weiand No. 8011 intake manifold and accompanying EP-15K camshaft are the hot ticket for increased low-end torque and snappy throttle response.

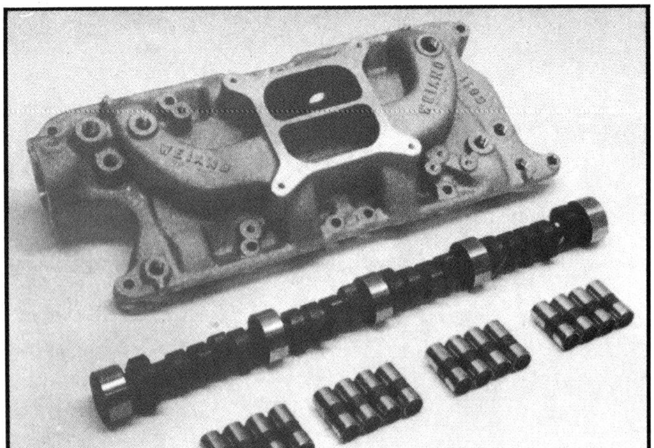

TEST 2

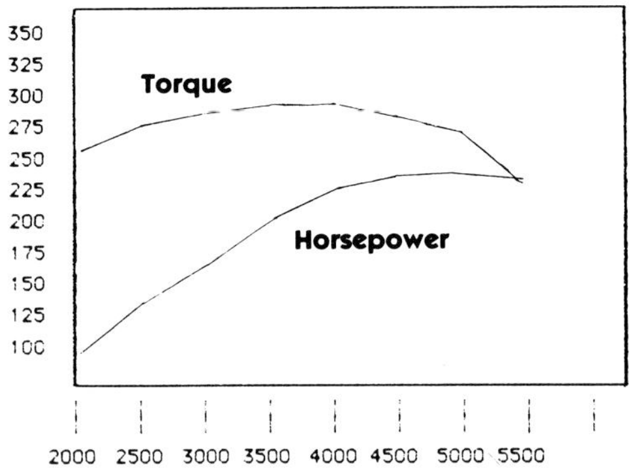

If you want your power higher in the rpm band, select the high-rise No. 8006 intake with EP-15P camshaft and full plenum divider. Both manifolds are available with EGR ports.

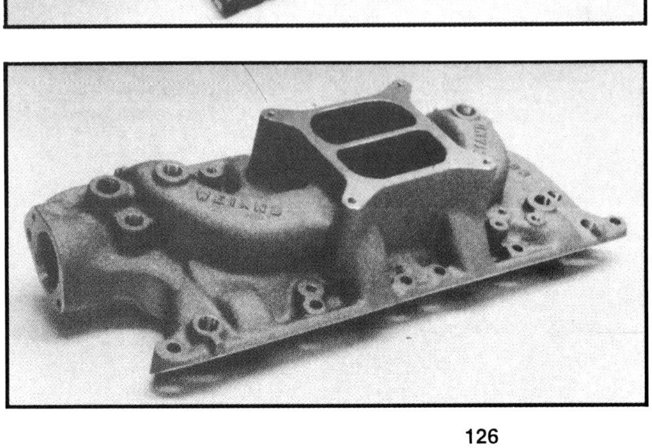

TEST 3

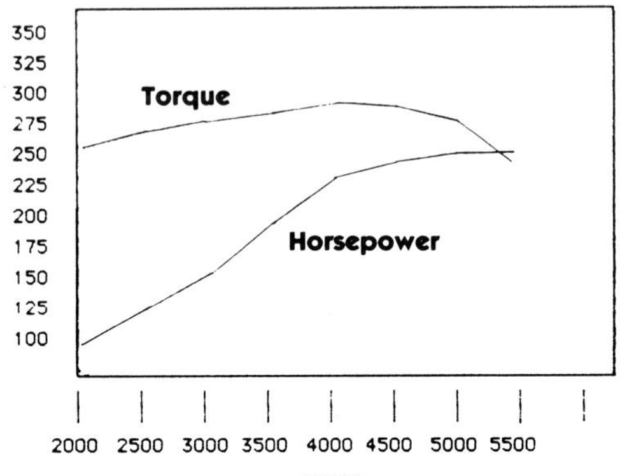

COMBOS

Intake manifold science has always fascinated hot rodders, probably because it's relatively easy and inexpensive to experiment with them. An intake that doesn't work on a particular combination is almost never a write-off because it can always be peddled to another rodder who might be able to use it. But while it's usually an interesting driveway exercise to try a new manifold on your car, the chances of meeting with spectacular success are only average. There's just no way that intake manufacturers can predict or even design for all the combinations of cylinder heads, camshafts, carburetors, compression ratios, etc. that rodders are going to run in their cars. Therefore, even the best manifold is almost always a compromise because you're going to use it with a variety of other components that can have a dramatic effect on its performance.

Seeking the ever-elusive compromise, intake manufacturers eventually identified the manifold/camshaft combination as one of the more significant relationships and began studying ways to make them complement one another. Weiand's Accelerator series is a familiar example. Their initial manifold/camshaft combinations for the small-block Chevy V8 were so successful, they have now followed up with a similar system for the small-block Ford V8. The new Ford combinations are based on two manifolds and two camshafts: one pair designed to improve low-speed torque and throttle response, the other designed to put some good high-speed power into an otherwise stock 302. The No. 8011 manifold is a low-profile unit that retains stock hood clearance while boosting your car's low end with help from Weiand's EP-15K hydraulic camshaft. The 8006 is a high-rise version that holds its own down low while really pumping up the power curve upstairs, thanks to a stronger EP-15P cam grind.

The accompanying dyno charts illustrate the dramatic difference that the manifold/camshaft combination is capable of producing in your small-block Ford engine. Corrected horsepower and torque numbers show a separate and distinct power increase in the targeted rpm range for each manifold and camshaft combination. These gains assume an otherwise stock engine, and it would be quite easy to spoil the effect by slapping on a big carburetor or headers with excessively large primary tubes. The improvements made available are intended to provide significant performance gains without misdirecting your engine's real application as a street driver. Keep that in mind when selecting other modifications for your car and you just might end up with a surprisingly strong street combination. **HR**

TEST 1:
Stock Baseline
Engine: Small-block Ford V8

Speed rpm	CBT lbs.-ft.	CBP hp	VE%	A/F	BSFC lbs./hphr
2000	248.9	94.8	83.8	12.4	.57
2500	265.3	126.3	87.9	13.5	.52
3000	276.9	158.2	89.7	12.0	.57
3500	283.1	188.7	90.5	12.2	.55
4000	278.7	212.3	91.0	12.4	.56
4500	262.3	224.7	88.4	12.3	.58
5000	239.2	227.7	85.0	12.6	.59
5500	211.2	221.2	78.7	12.2	.64

TEST 2:
Camshaft: EP-15L
Intake: No. 8011
Engine: Small-block Ford V8

Speed rpm	CBT lbs.-ft.	CBP hp	VE%	A/F	BSFC lbs./hphr
2000	257.2	97.9	88.3	14.8	.48
2500	276.2	131.5	89.1	15.1	.45
3000	284.1	162.3	91.5	13.3	.50
3500	294.6	196.3	93.6	13.4	.49
4000	295.2	224.8	96.3	12.2	.56
4500	282.1	241.7	94.4	12.5	.56
5000	256.4	244.1	89.7	12.4	.59
5500	227.0	237.7	84.9	12.2	.64

TEST 3:
Camshaft: EP-15K
Intake: No. 8006
Engine: Small-block Ford V8

Speed rpm	CBT lbs.-ft.	CBP hp	VE%	A/F	BSFC lbs./hphr
2000	257.3	98.0	87.5	12.4	.57
2500	265.2	126.2	89.0	12.0	.58
3000	274.5	156.8	93.0	11.5	.61
3500	289.8	193.1	95.7	12.4	.55
4000	296.2	225.6	99.2	12.4	.56
4500	286.9	245.7	97.6	12.1	.58
5000	265.9	253.1	93.7	11.9	.61
5500	238.4	249.7	87.0	11.8	.65

Legend: CBT = Corrected Brake Torque
CBP = Corrected Brake Power
VE% = Volumetric Efficiency
A/F = Air/Fuel Ratio
BSFC = Brake Specific Fuel Consumption

TIMING SPECS
Camshaft: EP-15L
Intake: No. 8011
Engine: Small-block Ford V8

Duration @ .000 Lift:	Intake 282°	Exhaust 282°
Duration @ .050 Lift:	Intake 208°	Exhaust 208°
Lift at Cam Lobe	Intake .280	Exhaust .280
Lift at Valve	Intake .420	Exhaust .420
Timing at .050 Lift:		
	Open	Close
Intake	-3° BTDC	31° ABDC
Exhaust	39° BTDC	-11° ATDC
Centerline 111°		

TIMING SPECS
Camshaft: EP-15K
Intake: No. 8006
Engine: Small-block Ford V8

Duration @ .000 Lift:	Intake 290°	Exhaust 300°
Duration @ .050 Lift:	Intake 214°	Exhaust 224°
Lift at Cam Lobe	Intake .296	Exhaust .311
Lift at Valve	Intake .444	Exhaust .466
Timing at .050 Lift:		
	Open	Close
Intake	-1° BTDC	35° ABDC
Exhaust	48° BTDC	-4° ATDC
Centerline 112°		

POWER FAST & EASY

FRUGAL FORD

By John Baechtel

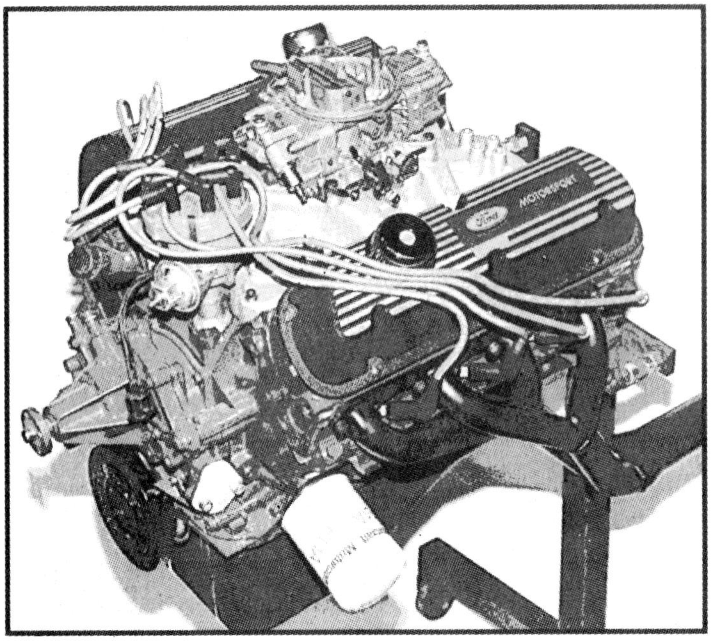

Single 4-barrel 302 or 351W small-block Fords easily achieve 300- to 400-hp, using off-the-shelf parts and standard engine-building procedures.

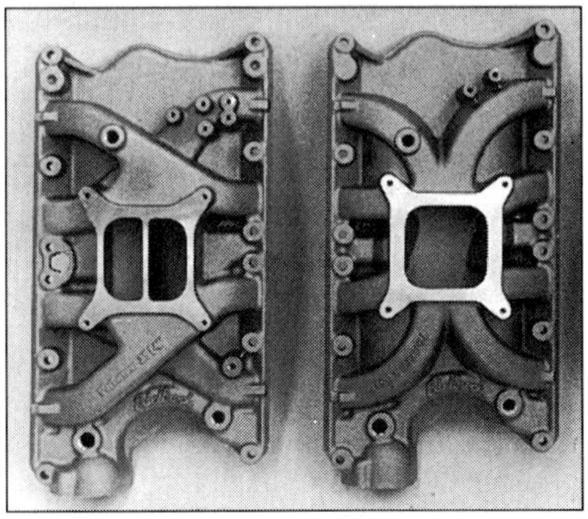

The dual-plane Edelbrock Performer (left) or single-plane Torker II intake manifolds are great choices for producing reliable horsepower. The dual-plane manifold will favor low-end torque, while the single-plane will give you more grunt at a higher rpm.

Building a powerful small-block Ford on a strict budget may seem like a formidable challenge, but it is still possible to achieve the magical one-hp-per-cubic-inch level and more. Just follow a few basic steps and select only the parts that will do the job and nothing more. Breathing is the key to success, and the Ford small-block is blessed with plenty of factory and aftermarket hardware that fills its cylinders to maximum capacity. By using a streetable 9.0:1 to 9.5:1 compression ratio, good cylinder heads, and proper inlet and exhaust systems, you can pull 300 to 400 horsepower from the Ford small-block V8 and maintain good driveability and reasonable fuel economy.

The nature of this article requires us to assume that you have an acceptable short-block with a strong, compatible crankshaft and connecting rods. Forged pistons are a plus, but four-bolt mains aren't really necessary. Your primary concerns are the components that affect breathing. Your greatest gain will come from a good set of cylinder heads. You can have a hot cam, a state-of-the-art intake manifold, and all the carburetors you want, but it's all for naught if the heads won't move air. The difference between the 300- and 400-hp versions is that the primary components regulate breathing. While 300 horsepower is easy to produce with either a 302 or 351, 400 horsepower is more difficult to obtain with any sort of streetable 302. Remember, the cylinder heads can make or break you.

If you're going to stick with factory heads, several choices are available. Presently, the hot setup is the iron GT-40 head (part No. M-6049-L302). It is a bolt-on replacement head for the 5.0L V8, and it accepts all standard valve gear. It carries 1.84-inch-diameter intake valves and 1.54-inch exhaust valves, both with hardened seats. The GT-40 significantly outflows the stock 5.0L heads and can be used successfully on 325-hp versions of the 5.7L V8.

The aluminum Windsor head (part No. M-6049-J302) is even better. This factory cylinder head holds 1.94-inch intake valves and 1.6-inch exhaust valves. It carries a 58cc chamber with 170cc intake ports. It works with both dual-plane and single-plane intake manifolds, but it has no provisions for manifold heating or EGR. You can expect good power from these heads, even in an unported form.

You also have a choice between the new World Products cylinder head and a variety of other performance cylinder heads, including ported factory heads that are offered by aftermarket sources. The Windsor head from World Products is an affordable bolt-on for both 302 and 351W engines. It features hardened-steel exhaust seats, 64cc chambers, 195cc venturi-shaped intake ports, raised spark plugs, 2.02-inch intake valves, and 1.6-inch exhaust valves. If you're aiming for 400-plus horsepower, this head will help you reach that mark.

Selecting a camshaft for these engines is relatively straightforward. You can get everything you need with a hydraulic cam. It might be a little easier to approach the 400-plus

302-351W FORD

	300 HP	400 HP
CARBURETOR	600 to 750 cfm	700 to 750 cfm
INTAKE MANIFOLD	Edelbrock Performer, Holley No. 300-6, SVO M-9424-A321 (302), SVO M-9424-A341 (351W)	Edelbrock RPM, Holley No. 300-18, SVO M-9424-B302 (302), SVO M-9424-A351 (351W)
CAMSHAFT (Hyd) Duration (@ .050)		
Intake	204°/204°	224°/226°
Exhaust	214°/216°	234°/234°
Lift Intake	.448/.456 in	.496/.491 in
Lift Exhaust	.472/.484 in	.520/.518 in
Lobe Separation	112°-114°	110°-112°
HEADERS		
Tube dia.	1⅝ - 1¾ in	1¾ in
Collector dia.	3 in	3½ in
HEADS		
Intake Valve dia.	1.84-1.94 in	1.94-2.02 in
Exhaust Valve dia.	1.54-1.60 in	1.60 in
Chamber size	58-64 cc	58-64 cc

Continued on page 132

By Rob Kinnan

To be a Ford lover is to be somewhat of a masochist. Paying $100 for the same basic part that goes for $50 in Chevy trim can really burn a guy, especially when the power production is equivalent. Add that to the notorious design changes and parts noninterchangeability consistent with Ford engines, and it's enough to keep all but the truly dedicated away from Henry's Dearborn-bred products. While building a righteous big- or small-block Ford certainly isn't as easy or inexpensive as some other brands, it can be done, and thankfully there are plenty of good parts out there to make it happen. That goes double when it comes to cylinder heads. Over the last year or so, some very good Ford heads have surfaced from both Dearborn and the aftermarket, so we thought we'd take a look at what's available. Let the drooling begin.

FORD CYLINDER-HEAD GUIDE

SMALL-BLOCK HEADS (289/302/351W)

The small-block is the staple of Ford performance parts. From the 289 to the 351 Windsor, a dramatic performance increase can be had from a good set of cylinder heads. Factory production small-block heads really choke these motors and keep them from making decent power without major porting. Tiny exhaust ports, barely adequate intake ports and small valve sizes are not conducive to horsepower, so swapping on a good pair of heads should be your first modification if real power is the goal.

FORD MOTORSPORT GT-40 CAST IRON

A bolt-on upgrade for late-model 5.0L engines, GT-40 iron heads also fit earlier Windsor-type small-blocks and have 1.84/1.54-inch valves, better-flowing ports, hardened exhaust seats, 65.6cc chambers (when used with Motorsport valves) and bolt-on rockers. They are available bare (part No. M-6049-L302) or complete (part No. M-6049-L303) and have the stamp of approval from the California Air Resources Board (CARB) with E.O. number D-308. In bare form, they retail for $359.95 each, $585.95 each complete.

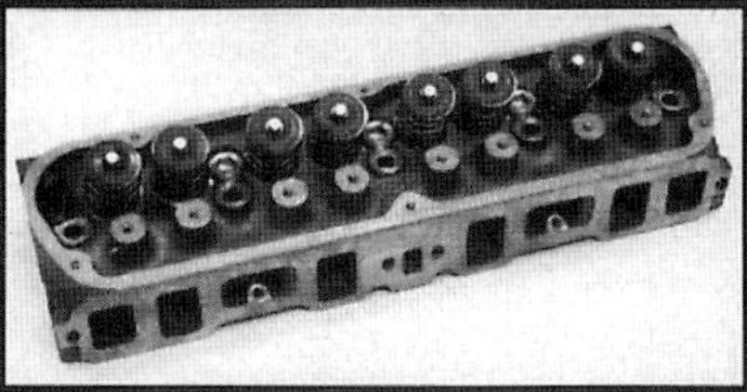

FORD MOTORSPORT GT-40 ALUMINUM

Ford SVO claims a 40-horse gain simply by bolting a pair of the new GT-40 aluminum heads on a late-model 5.0 Mustang, but they'll work just as well on earlier small-blocks. The intake and exhaust ports are bigger than stock for more flow potential, the 64cc chambers are a new "turbo-swirl" design, and the valve sizes are 1.94/1.54 inches. These heads also have more meat around the ports and on the deck surface to allow for more drastic porting or milling, and they're fully compatible with stock or Motorsport manifolds and headers. The assembled heads (part No. M-6049-Y303) come with stainless valves and dual valvesprings, which are compatible with all SVO roller camshafts. They are also available bare (part No. M-6049-Y302). Retail pricing hasn't been set in stone—make that aluminum—yet, but they should be affordable.

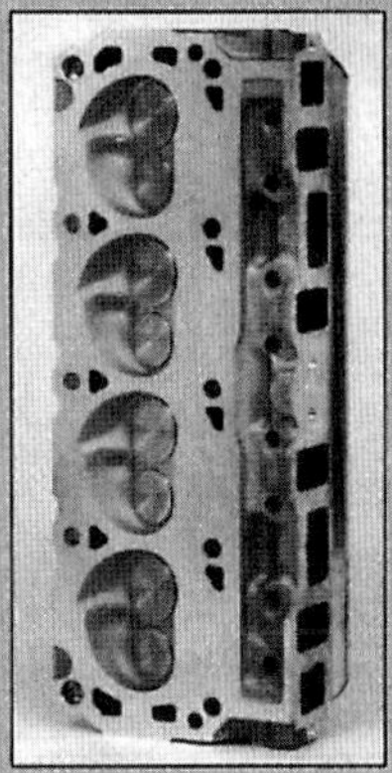

FORD MOTORSPORT WINDSOR BIG-PORT IRON RACING

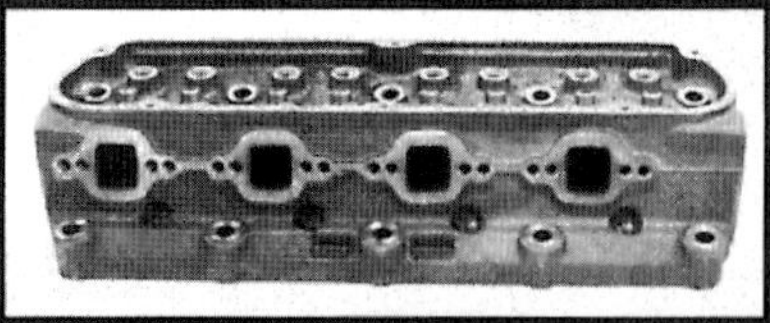

Dart cast these big-port small-block heads (part No. M-6049-E351) to Ford SVO's specs, and they're aimed at high-revving drag-racing applications or 351W short-track motors. Large 190cc intake ports, 2.02/1.60-inch valves and small 64cc combustion chambers spell out power in the higher rpms. They would be a good choice on a stout street 351, but they may be too much for all but the most serious 289/302. Other details include exhaust valve-seat inserts and stud-mounted rocker arms. The heads are sold only as bare castings and cost $395 each.

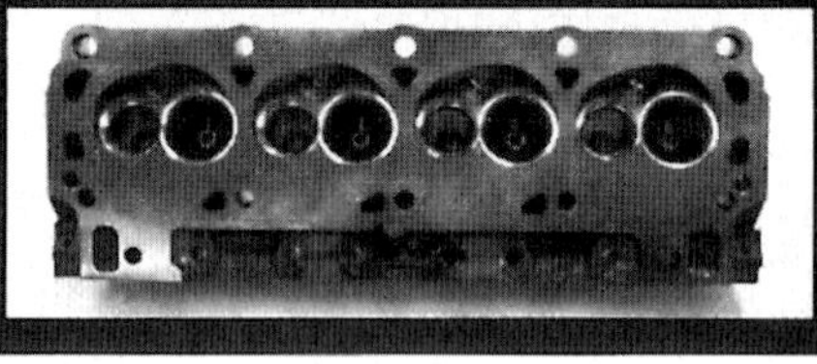

FORD MOTORSPORT SVO ALUMINUM HIGH-PORT NASCAR "YATES" HEADS

Ford's other hard-core, small-block cylinder-head offering was designed by Winston Cup guru Robert Yates and is the standard-issue Ford Winston Cup head. That means it's very good, although probably a little "out there" for a street engine due to the 210- to 228cc intake ports and a minimum intake valve size of 2.100 inches. Here, it's shown on the left next to the previous C302 head that it originally started life as. The chart below shows the different combinations available with this head. The C33 version is CNC ported. It's high-digit, but it's the small-block head to have if it's no-compromise time.

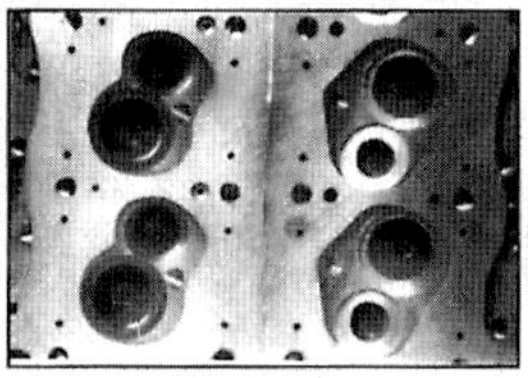

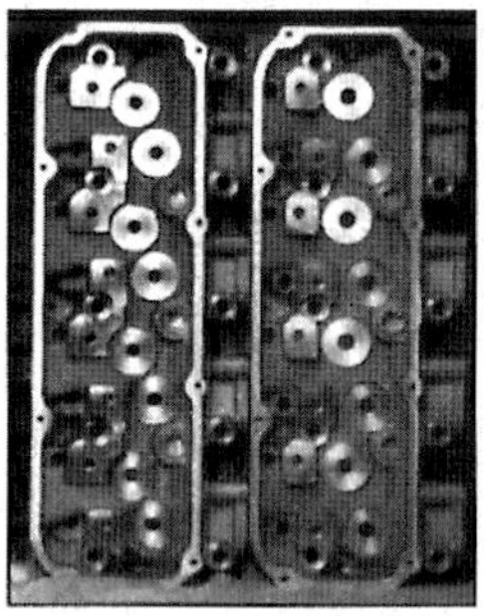

		Port Volume		Valve Diameter		Price ($)
Part No.	Chamber cc	Intake	Exhaust	Intake	Exhaust	
M-6049-C3	40.0	210 cc	119 cc	2.100	1.600	1225.95
M-6049-C3L	67.0	210 cc	119 cc	2.100	1.600	1225.95
M-6049-C302 -C302B	65.0	212 cc	95 cc	2.150	1.650	622.00
M-6049-C33	48.0	228 cc	131 cc	2.150	1.650	2640.95

AR INCORPORATED ALUMINUM

Alan Root's small-block head used to have Ford part No. M-6049-J302. Ford recently discontinued this head from the catalog when the aluminum GT-40 head was ready to come to market, but AR still offers the "J302" head direct. Compared to a stock head, the J302 has bigger and better flowing ports (170cc intake, 60cc exhaust), 2.02/1.60 valves, a 58cc chamber and a stud-mounted rocker setup. These heads accept stock intake and exhaust manifolds and valve covers and are a bolt-on part, but they're not 50-state legal if that's a consideration. In bare form, AR sells them for $995, and a complete assembly with good springs goes out the door for $1395.

TRICK FLOW SPECIALTIES/SUMMIT

Trick Flow Specialties has two different heads for the small-block Ford, one in aluminum and the other made of cast iron, and they're both good for about 40 bolt-on horses. The iron head has CARB E.O. number D-216-5, making it smog-legal in all 50 states. The aluminum head has all the smog hookups necessary, but so far it hasn't been given an E.O. number. Both heads have much better ports than stock, relocated valves, raised exhaust ports and 60- or 69cc chambers, and the aluminum head has a thicker .600-inch-minimum deck surface. Standard valve sizes are 1.94-inch intake, 1.60-inch exhaust, but these can be increased to 2.08/1.65 on the aluminum head and 2.125/1.65 on the iron head. And the price is reasonable, too. Summit Racing sells the iron heads for $695 bare, $1195 assembled ($775/$1275 for the smaller chambers), and the aluminum units are $1295 bare, $1695 assembled ($1375/$1775 for 60cc chambers).

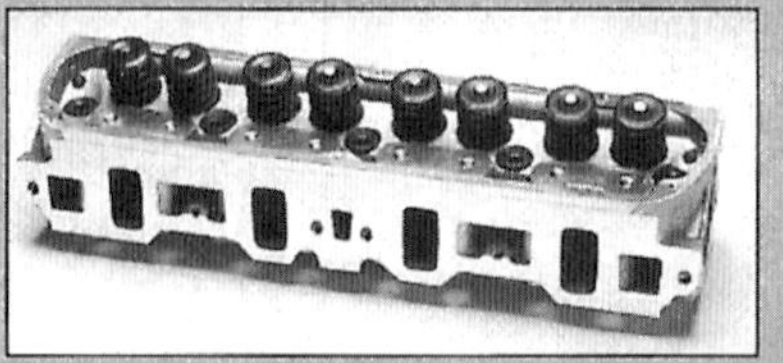

EDELBROCK

Edelbrock has three different twists on the same small-block Ford aluminum head, depending on your application. The Performer, Performer 5.0 and Performer RPM all share the following features: aluminum construction, 170cc intake runners, 60cc exhaust runners, 60cc chambers, CNC porting, a 5⁄8-inch-thick deck surface, 1.60-inch exhaust valves, 1.450-inch-diameter springs and availability in bare castings or fully assembled heads. The differences are that the standard Performer heads have emissions provisions and exhaust crossovers, making them 50-state legal, whereas the RPM heads don't have these provisions. Both of these heads use stud-mounted rocker arms, screw-in studs and guideplates. The Performer 5.0 head is a direct bolt-on for '82-'92 5.0s and uses the stock pedestal-mount rocker arms and valvetrain. The Performer and Performer RPM are also available with either 1.90- or 2.02-inch intake valves, but the 5.0 head is only available with 1.90s. The smaller valves allow the head to bolt on without cutting valve reliefs in the pistons, which may be necessary with the 2.02 valves. The retail price for the RPM head is $1375.95 assembled, and the others are $1449.95, but we've seen them advertised for less than $1000.

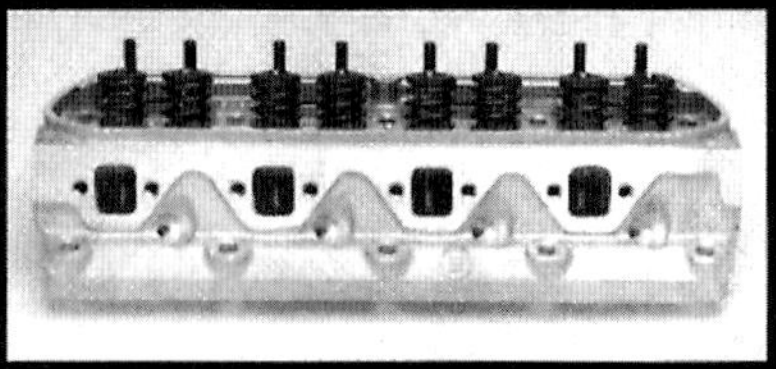

WORLD PRODUCTS WINDSOR/WINDSOR JR.

Instead of offering a single part to fit a variety of engine sizes, World Products has two small-block Ford heads that fit specific-size engines and applications. The Windsor is the bigger of the two, with 197cc intake ports, 2.02/1.60 valves and 64cc combustion chambers. The other head is known as the Windsor Jr., and it comes with smaller 171cc intake ports, 1.94/1.60 valves and 58cc chambers. Both heads are available bare or fully assembled in aluminum or cast iron, and both have revised intake and exhaust runners to improve flow without losing the all-important velocity. But the ports are in the stock locations, so stock intake and exhaust manifolds bolt up perfectly. The most radical improvement over stock heads are the World Products exhaust ports, which are significantly bigger than the weak-kneed production heads. Both the Windsor and the Windsor Jr. have dual-exhaust-manifold bolt patterns—one for standard manifolds and another for big-tube headers. Another trick feature that both heads share is an adjustable valvetrain that allows more aggressive camshaft timing; stock 289 rockers, standard stamped rockers or rollers can be used. These heads are 50-state-legal bolt-ons.

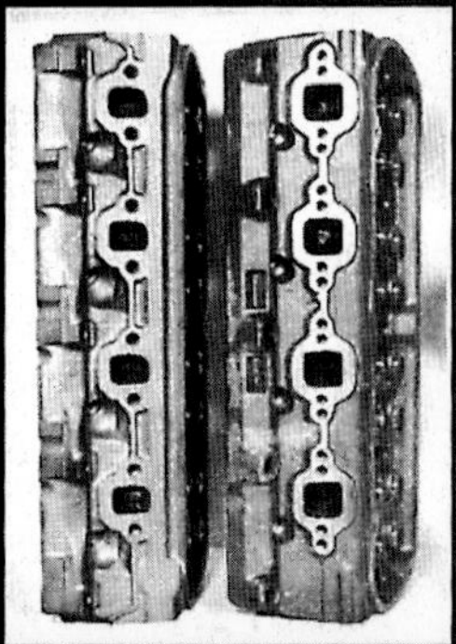

Which head should you use? That depends on engine displacement and configuration. World Products recommends that nearly all 351W engines use the bigger Windsor. High-compression, high-winding 289s can get away with the Windsor, but anything less should use the Jr. The 302 is a little more forgiving. The Jr. should be used on a 302 with a stock cam and intake, but the Windsor is better suited when more compression, cam timing or intake volume come into play. When a Windsor is used on a 289 or 302, the larger chambers will lower the compression ratio, so to get the ratio back where it should be, either mill the heads or get pistons with more dome volume. The price for power is $645 bare, $945 assembled for the iron heads and $1198 and $1498 for the aluminum versions. There are also several options ranging from mild pocket porting to fully ported and assembled heads for $2239.

BIG-BLOCK HEADS (429/460)

Tired of messing around with small-blocks? Want some real cubes under the hood? Then build a 429 or 460 Ford. Or bigger. Speed parts for the 429/460 aren't as common as for a small-block, but there are a few good cylinder heads for the big Fords. And things such as camshafts and pistons are plentiful and will make serious horsepower if matched with the correct heads for the application. When it comes to trick big-block Ford cylinder heads, you basically have three choices. Which choice you make is dictated by how extreme the engine is.

FORD MOTORSPORT COBRA JET ALUMINUM HEADS

The Cobra Jet heads (part No. M-6049-A429) were designed by Alan Root (AR Incorporated) and Jon Kaase, are manufactured for Ford by AR and are dynamite street big-block cylinder heads. In fact, they're the only aftermarket big-block heads you can realistically use on a "real" street car. The CJ is a bolt-on and allows the use of factory-type intake and exhaust manifolds and a production valvetrain. Details include 307cc intake ports, 130cc exhausts, 2.250/1.750-inch valves and 77cc chambers. And at 28 pounds apiece, the head weighs less than half of what a stock iron head weighs. In the April '92 issue, HOT ROD reported on a dyno test on the CJ heads. On a stock 460, the heads alone were worth 104 horsepower and 43 lbs-ft of torque (460 and 488 respectively) with increases in both across the board. With more cam, intake and carb, the numbers went to 650 horsepower and 542 lbs-ft. When used with flattop pistons, these heads give a 10.5:1 compression ratio. The retail price for the CJ head is $842.95 each.

Continued from page 128

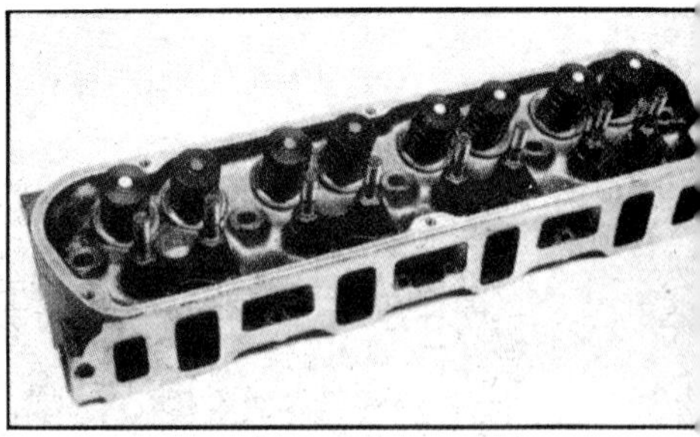

World Products' Windsor heads use 64cc chambers, a raised port roof, standard Chevy 2.02-inch intake valves, and 1.60-inch exhaust valves to produce excellent power on any small-block Ford.

horsepower mark with a mechanical cam, and, of course, a street roller is the answer if money is no object. For a smooth idling engine with great low-end torque, look for a hydraulic cam with a 112- to 114-degree lobe center separation, a .450- to .480-inch valve lift and 210 to 224 degrees duration at a .050-inch valve lift. If you want to pull some serious power out of the engine, you'll have to live with a rougher idle and reduced low-speed torque, but the midrange and top-end power surge should quickly make up for any lack of low-speed power. Look for a hydraulic or mechanical cam with a 110- to 112-degree lobe separation, .480- to .520-inch lift, and 224 to 244 degrees of duration at a .050-inch lift. Remember to check the piston-to-valve clearance with these larger cams, and also check for things such as the retainer-to-guide clearance and valvespring coil bind.

Intake manifold choices are easy. You can't go wrong with an Edelbrock Performer or RPM. Holley offers dual-plane and single-plane manifolds that will produce the power you want, and the various SVO intakes will also deliver strong power. Keep carburetor size down to around 600 to 650 cfm on the 302, unless you're building a lot of power. The 351 doesn't need more than 750 cfm, even at the 400-plus-horsepower level. While a Holley vacuum secondary is generally the hot ticket, a double-pumper can be used.

Keep the header size in line with the rest of your combination and run it through a free-flowing exhaust system for the best results. As always, make certain you give the engine adequate ignition energy and a compatible spark curve. Use a high-flow air filter element and spend some time optimizing the combination at the track. The results will not only surprise you, but they'll be easy to live with on the street. **HR**

FORD MOTORSPORT PRO STOCK HEADS

A big step up in performance from the Cobra Jet heads are the SVO 460 Pro Stock aluminum heads. These are manufactured for Ford by AR and have raised intake and exhaust ports, altered valve angles (7.5 degrees intake, 8.0 degrees exhaust), 65cc wedge-style combustion chambers and enough runner volume and flow potential to feed the hungriest of motors. The port and chamber configurations are based heavily on the Robert Yates-designed SVO 351 heads, so they are also commonly referred to as "Yates heads." There are actually two versions of these heads available: one with a standard head bolt pattern (part No. M-6049-C460, $797.50 each bare) and one with six bolts per cylinder (part No. M-6049-D460), which requires bolt-on lugs in the lifter valley of the block and relocated bolt holes in the block. The recommended minimum bore size when using these heads is 4.600 inches, and the standard valve sizes are 2.400 inches intake, 1.850 inches exhaust. A stock valvetrain won't work on the Yates heads—an individual rocker-shaft valvetrain, such as a Jesel setup, is required. Also, Ford only has one intake manifold to fit these heads, and it's a dual-plane single-four-barrel intake (part No. M-9424-C460). Because of the physical size of the heads, they would be a nightmare-fit in a stock shock-tower-equipped Ford engine compartment. So don't plan to bolt these onto a 460 and throw it into a stock Fairlane or Mustang—it won't fit! These are serious parts, but they would work well on a Fastest Street Car-style motor, and ya know, we haven't seen a big-block Ford at Memphis yet.

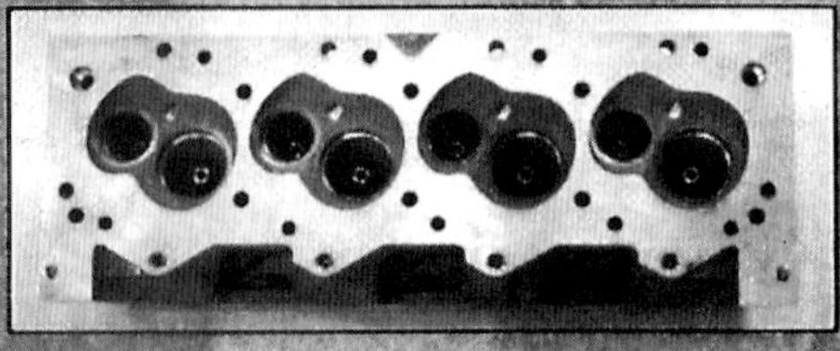

AR BOSS 429 HEMI

Designed only for full-tilt professional competition engines, the Alan Root 429 Hemi heads are at the top of the heap in big-block Ford performance cylinder heads. These are the ones to use in Pro Stock and Pro Mod racing, and they will accept up to manhole-cover-size 2.700-inch intake valves and 2.100-inch exhaust valves. The chambers measure 85 cc, and the ports will flow enough air to feed a mile-long line of locomotives. A fully ported head will flow an amazing 570 cfm on the intake side! The Hemi heads use a standard bolt pattern to fit a stock-type 460 block, an AR or Jesel rocker system must be used, and AR makes the only intake manifolds to fit. One is a cast-aluminum tunnel-ram, and the other is a blower manifold, but most racers use custom sheetmetal intakes. Not for the faint of heart, these heads cost $2800 a pair in bare form. **HR**

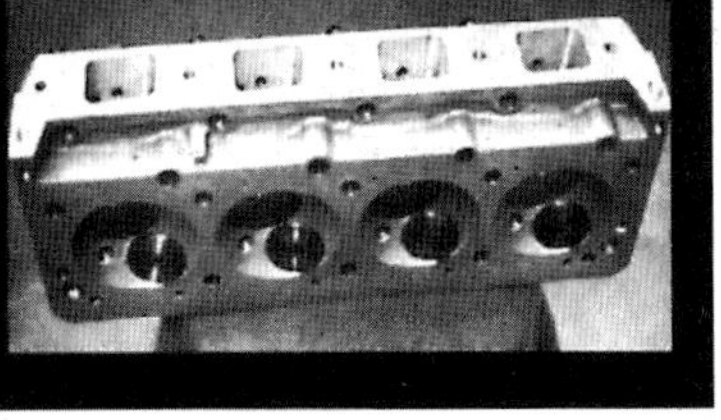

SOURCES

AR Incorporated
Dept. HR09
1696 Walter St.
Ventura, CA 93003
805/654-0654

Dart Machinery
Dept. HR09
353 Oliver St.
Troy, MI 48084
313/362-1188

Edelbrock Corporation
Dept. HR09
2700 California St.
Torrance, CA 90503
310/781-2222

Ford Motorsport/SVO
Dept. HR09
44020 N. Groesbeck Hwy.
Mt. Clemens, MI 48083-1115
313/337-1356

Summit Racing Equipment
Dept. HR09
P. O. Box 909
Akron, OH 44309
216/630-0200

Trick Flow Specialties
Dept. HR09
1248 Southeast Ave.
Tallmadge, OH 44278
216/630-1555

World Products
Dept. HR09
35 Trade Zone Dr.
Ronkonkoma, NY 11779
516/737-0372